The Luoping Biota:
A taphonomic window on Triassic biotic recovery and radiation

罗平生物群
——三叠纪海洋生态系统复苏和生物辐射的见证

编委会名单

胡世学　　　张启跃　　　文　芝　　　黄金元
周长勇　　　谢　韬　　　吕　涛
**Shixue Hu　　Qiyue Zhang　　Wen Wen　　Jinyuan Huang
Changyong Zhou　Tao Xie　　Tao Lü**
中国地质调查局成都地质调查中心
Chengdu Center of China Geological Survey

刘腾林
Tenglin Liu
罗平县国土局
Land and Resources Bureau of Luoping County

迈克尔·本顿
Michael J. Benton
英国布里斯托大学
University of Bristol

罗平生物群
The Luoping Biota:
——三叠纪海洋生态系统复苏和生物辐射的见证
A taphonomic window on Triassic biotic recovery and radiation

国家出版基金项目
NATIONAL PUBLICATION FOUNDATION

云南出版集团公司
云南科技出版社
·昆明·

YUNNAN PUBLISHING GROUP CORPORATION
YUNNAN SCIENCE & TECHNOLOGY PRESS
·KUNMING·

图书在版编目（CIP）数据

罗平生物群 / 胡世学等著. -- 昆明：云南科技出版社，2016.9
ISBN 978-7-5587-0110-8

Ⅰ.①罗… Ⅱ.①胡… Ⅲ.①生物群—古生物学—研究—罗平县 Ⅳ.①Q911.727.44

中国版本图书馆CIP数据核字(2016)第232935号

责任编辑：杨旭恒　赵　敏　章　沁
整体设计：晓　晴
责任校对：叶水金　张彦艳
责任印制：翟　苑

云南出版集团公司
云南科技出版社出版发行
（昆明市环城西路609号云南新闻出版大楼　邮政编码：650034）
昆明富新春彩色印务有限公司印刷　全国新华书店经销
开本：787mm×1094mm　1/12　印张：13　字数：240千字
2016年12月第1版　2019年7月第2次印刷
印数：1~3000册　定价：128.00元

罗平生物群——第三次生物大辐射的典型代表

殷鸿福 2015.2.28

"The Luoping biota: a typical example of the great third radiation in earth history", autography by academician Professor Hongfu Yin

前言

三叠纪是地球生命演化发展史上一个关键的时期。它开始于一场空前绝后的生物大绝灭，终结于另一场大规模的生物灭绝事件。二叠纪末期的生物大绝灭导致了陆地上70%的物种和海洋里90%以上物种的绝灭，彻底改变了地球的生态环境。三叠纪早期的海洋环境极不稳定，出现了全球范围的缺氧事件，赤道地区还经历了极端的高温环境。严酷的环境迟滞了生物的复苏，整个海洋生态系统直到三叠纪中期才从二叠纪末大绝灭的灾难性后果中全面复苏过来。

罗平生物群是中国地质调查局成都地质调查中心2007年进行1∶50000区域地质调查过程中首次发现并命名。经过成都地质调查中心罗平生物群研究团队多年的潜心研究，初步揭示了罗平生物群的面貌和科学意义。目前罗平生物群已发现的化石有6个门类40属100多种，其中大部分为新属新种。罗平生物群种类丰富、保存精美，堪称世界级的化石宝库，被誉为三叠纪海洋生态系统全面复苏的代表，是海生爬行类、新鳍鱼类和甲壳类节肢动物辐射的窗口，是中生代海洋生态系统形成的标志。

本书图文并茂，大量精美的化石照片展示了栩栩如生的罗平生物群各门类化石，为国内外科学研究人员和科学爱好者了解罗平生物群、探索两亿多年前神秘的三叠纪海洋世界提供了一个难得的机会。

罗平生物群的研究得到了中国地质调查局及罗平县地方政府的大力支持。殷鸿福院士一直关注罗平生物群的研究并给本书题词。Brian Choo博士和昆明学院陈庆韬老师绘制了精美的化石复原图。中科院北京脊椎动物和古人类研究所徐光辉博士和中国地质调查局武汉地质调查中心程龙博士提供了部分图片。中国地质大学童金南教授、张克信教授、陈中强教授、北京大学江大勇教授、合肥工业大学刘俊博士、美国肯特州立大学Rodney M. Feldann教授、Carrie E. Schweitzer教授、德国柏林自由大学Helmut Keupp教授及Michael Steiner博士、西澳大学罗茂博士等对罗平生物群研究提供了大量帮助，在此一并致谢！

Foreword

The Triassic was a crucial period in the history of the evolution of life. It started after the severest mass extinction (the Permo-Triassic mass extinction) and ended by another mass extinction, and marked the remarkable replacement of Palaeozoic ecosystems by modern ecosystems. The Permo-Triassic mass extinction led to the loss of about 70% of terrestrial species and >90% of marine species globally. The unstable environment in the Early Triassic, marked by global anoxia and high temperatures in the equatorial region, hampered the recovery process significantly. It was widely accepted that the full recovery of marine ecosystems took nearly 8-10 million years, by the time of the Middle Triassic.

The Luoping Biota was discovered in 2007 by the Chengdu Center of the China Geological Survey during regional mapping work at the scale of 1:50,000. Several big excavations carried out in the following years revealed the fossil assemblage and the scientific significance. So far, 6 animal phyla, including more than 100 species and 40 genera have been recognized, most of which are new taxa. The exceptional preservation and high diversity of the fossils elevates the Luoping Biota to be counted as one of the most significant Lagerstätten in the world. The Luoping biota is seen as one of the most diverse Triassic marine fossil lagerstätten in the world, and it has been used as a marker of the recovery of marine ecosystems after the Permo-Triassic mass extinction. The Luoping biota also documents the diversification of crustacean arthropods, neopterygian fishes, and marine reptiles, and furthermore, the establishment of Mesozoic marine ecosystems by the Middle Triassic.

This book provides readers with hundreds of colourful pictures of the spectacular fossils from the Luoping biota. It offers a window into a true Middle Triassic community 244 million years ago. It also offers important lessons regarding the early evolution of Mesozoic marine ecosystems.

Research work on the Luoping biota was financially supported by the China Geological Survey. The field work was supported by the local government of Luoping County. We sincerely thank academician Professor Hongfu Yin for his help and encouragement over the years, and the autography for this book. We would like to thank Dr. Brian Choo and Mr. Qingtao Chen for their extraordinary reconstruction of the fossils, and Dr. Guanghui Xu (Institute of Vertebrate Paleontology and Paleoanthropology, Chinese Academy of Sciences), Dr. Long Cheng (Wuhan Center of China Geological Survey) for providing some important pictures. Thanks are also given to Prof. Jinnnan Tong, Zhong-qiang Chen, and Kexing Zhang (all from China University of Geosciences, Wuhan), Prof. Dayong Jiang (Peiking University), Dr. Jun Liu (Hefei University of Technology), Prof. Rodney M. Feldann and Dr. Carrie E. Schweitzer (both from Kent State University, U. S. A.), Prof. Helmut Keupp and Dr. Michael Steiner (both from Freie Universität Berlin, Germany), and Dr. Mao Luo (The University of Western Australia) for constructive and informative discussions.

The Luoping Biota
A taphonomic window on Triassic biotic recovery and radiation

目　录

罗平生物群的发现 / 1
Discovery of the Luoping Biota

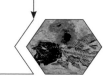

罗平生物群产地、层位及时代 / 5
Fossil sites, stratigraphy and age of the Luoping Biota

罗平生物群的特异埋藏 / 18
Exceptional preservation of the Luoping Biota

罗平生物群的科学意义 / 25
Perspectives of the Luoping Biota

CONTENTS

罗平生物群化石 / 28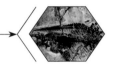
Fossils of the Luoping Biota

海生爬行类 Marine reptiles / 28

鱼类 Fishes / 43

节肢动物 Arthropods / 86

软体动物 Molluscs / 118

腕足动物 Brachiopods / 124

海绵动物 Sponges / 125

植物 Plants / 126

遗迹化石 Trace fossils / 128

罗平生物群国家地质公园 / 136
Luoping Biota National Geopark

主要参考文献 Selected references / 145

The Luoping Biota
A taphonomic window on Triassic biotic recovery and radiation

罗平生物群的发现
Discovery of the Luoping Biota

2007年10月，中国地质调查局成都地质调查中心云南1：50000区域地质调查项目组在野外地质调查过程中，在距罗平县城东南15km的罗雄镇大洼子村附近中三叠统关岭组二段地层中发现大量保存完好的鱼类化石，随后又发现了大量爬行类及其他动、植物化石。经过研究证实其生物门类的多样性、化石保存的完整性举世罕见，将其正式命名为"罗平生物群"。

The Luoping Biota was discovered in 2007 by the Chengdu Center of the China Geological Survey during regional mapping work at the scale of 1:50,000 from the Dawazi Village, Luoxiong Town, 12km southeast of the City of Luoping, Yunnan Province, SW China. Although fishes were the first group to be recovered, later excavation has led to the discovery of more marine reptiles and many other fossil groups. Later, in 2008, the unusual fossil assemblage was named as "The Luoping Biota".

2 | 罗平生物群
——三叠纪海洋生态系统复苏和生物辐射的见证
The Luoping Biota: A taphonomic window on Triassic biotic recovery and radiation

罗平县大洼子村——罗平生物群化石最早发现地
The Dawazi Village near the county town of Luoping, from where the Luoping fossils were recovered for the first time.

4 | 罗平生物群
——三叠纪海洋生态系统复苏和生物辐射的见证
The Luoping Biota: A taphonomic window on Triassic biotic recovery and radiation

最先发现的第一块罗平生物群化石,比例尺为1cm
The first recovered specimen of the Luoping biota. Scale bar is 1cm.

罗平生物群产地、层位及时代
Fossil sites, stratigraphy and age of the Luoping Biota

罗平生物群产地主要分布在罗平县城周围，主要化石点有大洼子、九光等地。最近在相邻的泸西、丘北县也发现了罗平生物群化石。目前在罗平进行了三次大规模的化石发掘，采集了大量化石，发掘所留的采场构成了罗平生物群国家地质公园的核心部分和主要景观。产罗平生物群化石的地层为中三叠世关岭组二段，根据含化石地层中发现的微体化石牙形石的鉴定结果，确认罗平生物群时代为中三叠世安尼期派尔逊亚期，从化石层相伴产出的火山灰夹层中锆石同位素测年获绝对年龄为距今约2.44亿年。

Fossil localities of the Luoping biota are recovered from the area surrounding the county town, among which Dawazi and Jiuguang are the two most important localities. Recently, similar fossils are also recovered from the neighboring Luxi and Qiubei Counties. Three big excavations have been carried out and hundreds of fossils have been collected. The three quarries from the excavations comprise the central part of the Luoping Biota National Geopark. The interval containing the Luoping biota is the middle part of Member II of the Guanling Formation. The age of the Luoping biota is assigned to the Pelsonian Substage of the Middle Triassic Anisian Stage based on the index conodont fossils. An age of 244 million years is estimated for the Luoping biota, based on zircons derived from volcanic tuff layers.

F 罗平生物群产地、层位及时代
Fossil sites, stratigraphy and age of the Luoping Biota | 7

九光村，另一个罗平生物群重要化石产地
Jiuguang Village, another important fossil locality of the Luoping biota.

8 | 罗平生物群
——三叠纪海洋生态系统复苏和生物辐射的见证
The Luoping Biota: A taphonomic window on Triassic biotic recovery and radiation

罗平生物群上石坎采场
Shangshikan quarry for fossil excavations.

关岭组二段中部产罗平生物群层位，约20m厚
The interval containing the Luoping biota from the middle part of Member II, Guanling Formation, approximately 20m in thickness.

10 罗平生物群
——三叠纪海洋生态系统复苏和生物辐射的见证
The Luoping Biota: A taphonomic window on Triassic biotic recovery and radiation

上石坎采场全景
Panorama of the Shangshikan Quarry.

罗平生物群
——三叠纪海洋生态系统复苏和生物辐射的见证
The Luoping Biota: A taphonomic window on Triassic biotic recovery and radiation

古生物化石和喀斯特地貌、油菜花海的完美结合
A fascinating fossil site located in a wonderful landscape of karst mountains and canola fields.

响洞坡采场
Xiangdongpo quarry for fossil excavations.

上石坎采场远眺
Overview of the karst landscape from the top of the Shangshikan quarry.

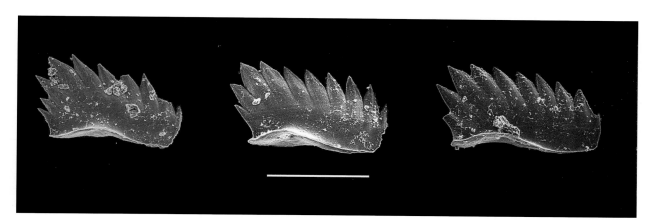

科克尼尼古拉刺，中三叠世安尼期标准化石，比例尺为 200 μm

Nicoraella kockeli, the index fossil of the Pelsonian substage, Middle Triassic. Scale bar is 200 μm.

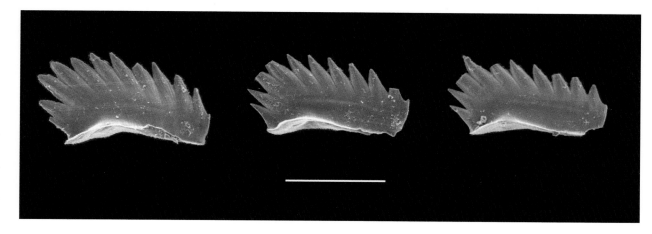

德国尼古拉刺，中三叠世安尼期标准化石，比例尺为 200 μm

Nicoraella germanicus, the index fossil of the Pelsonian substage, Middle Triassic. Scale bar is 200 μm.

2009年5月，上石坎采场化石发掘
Field excavation at Shangshikan Quarry, May 2009.

工作人员在显微镜下修理化石
Preparation of fossils under a microscope.

罗平生物群的特异埋藏
Exceptional preservation of the Luoping Biota

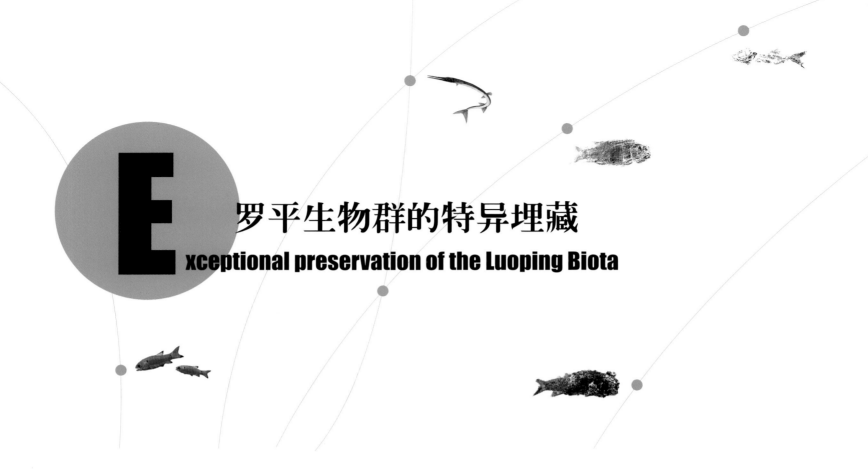

　　罗平生物群化石保存精美，属于典型的特异埋藏。特异埋藏的化石被称为"化石记录的珍珠"，在古生物学研究上具有独特的意义。特异埋藏的化石除了能够保存通常情况下难以保存的完整的骨骼化石外，许多弱矿化的生物体及生物软体构造也得以完好保存，对研究生物演化以及地质历史时期生态系统变迁具有重要意义。

　　产罗平生物群化石的地层为中三叠统关岭组二段中部，岩性为泥晶灰岩与泥灰岩。从保存上看，罗平生物群属于Platenkalk类型。罗平生物群的埋藏环境为台内坳陷，水体底部处于缺氧环境。保存化石的层面上常见微生物席，说明罗平生物群化石的特异埋藏与微生物席的密封作用有关。大量保存完好的化石出现在单个层面上，说明生物系灾难性造成的集群绝灭。藻类等生物大量繁盛造成的缺氧和中毒事件可能是罗平生物群生物死亡的主要原因。

Exceptional fossils, known as "prizes of the fossil record", are crucial in our understanding of life evolution and ecosystem dynamics by the preservation of non-mineralized and slightly mineralized tissues, otherwise the preservation potential of soft tissues is extremely low in normal taphonomic condition.

The intervals containing the exceptionally preserved Luoping fossils belong to the middle part of Member II of the Guanling Formation. The rocks are mainly micritic limestones and bituminous shales. The preservation of the Luoping biota is similar to the Mesozoic Plattenkalk from South Germany, which is characteristic by paired beds of limestone and mudstone. The depositional environment of the fossil layers is interpreted as an intraplatforrm basin, with stagnant and anoxic bottom waters. Microbially induced sedimentary structures are commonly observed from the bedding surface of the fossil layers, indicating that sealing by microbial mats might have contributed to the preservation of the exceptional fossils. The mass occurrence of single taxa on certain layers indicates a mass killing of swarming animal groups. A process of eutrophication, in which the water is starved of oxygen, resulting from an algal bloom, may have been the cause.

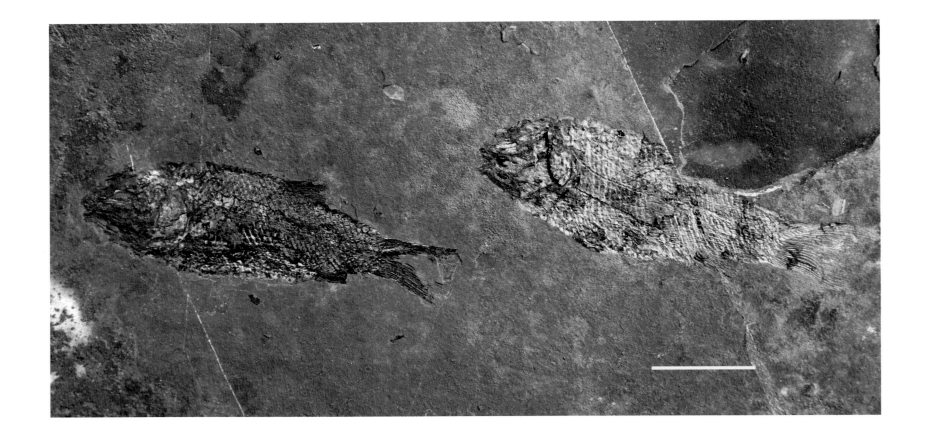

刚刚暴露在空气中的鱼化石，比例尺为1cm
Fresh sample of fishes immediately after excavation. Scale bar is 1cm.

经历过风化的鱼化石，比例尺为1cm
Fossil fish underwent weathering. Scale bar is 1cm.

保存在微生物席上的龙鱼化石，比例尺为5cm
Fossil saurichthyid preserved on the surface with microbial mats. scale bar is 5cm.

群体保存的龙虾化石，比例尺为1cm
Cluster of lobsters. Scale bar is 1cm.

罗平生物群
——三叠纪海洋生态系统复苏和生物辐射的见证
The Luoping Biota: A taphonomic window on Triassic biotic recovery and radiation

在层面上密集保存的鱼化石，比例尺为1cm
Cluster of fishes on the bedding surface. Scale bar is 1cm.

罗平生物群的科学意义
Perspectives of the Luoping Biota

罗平生物群目前已经发现有海生爬行类、鱼类、节肢动物、棘皮动物、菊石、双壳、腹足类、腕足类、植物等10多个大类的化石，完好地体现了当时海洋生物的多样性，是目前已知化石分异度最高的三叠纪海生化石库之一。罗平生物群是二叠纪末生物大灭绝后海洋生态系统的全面复苏的代表，是中三叠世生物大辐射的窗口，是中生代海洋生态系统形成的标志。罗平生物群保存精美，种类丰富，规模庞大，堪称世界级的化石宝库。

The fossil assemblage of the Luoping biota is a mixture of marine animals, terrestrial plants and a few terrestrial animals. To date, more than ten fossil groups have been recovered, including marine reptiles, fishes, arthropods, echinoderms, ammonites, bivalves, gastropods, lingulid brachiopods, foraminifers, and plants. The Luoping Biota is one of the most diverse Triassic marine fossil Lagerstätten records in the world, showing the full recovery of marine ecosystems after the end-Permian mass extinction. It also documents the diversification of crustacean arthropods, neopterygian fishes, and marine reptiles, and furthermore, the establishment of Mesozoic marine ecosystems by the Middle Triassic. The exceptional preservation and high diversity of the fossils elevates the Luoping biota to be counted as one of the most significant Lagerstätten in the world.

罗平生物群的科学意义
Perspectives of the Luoping Biota

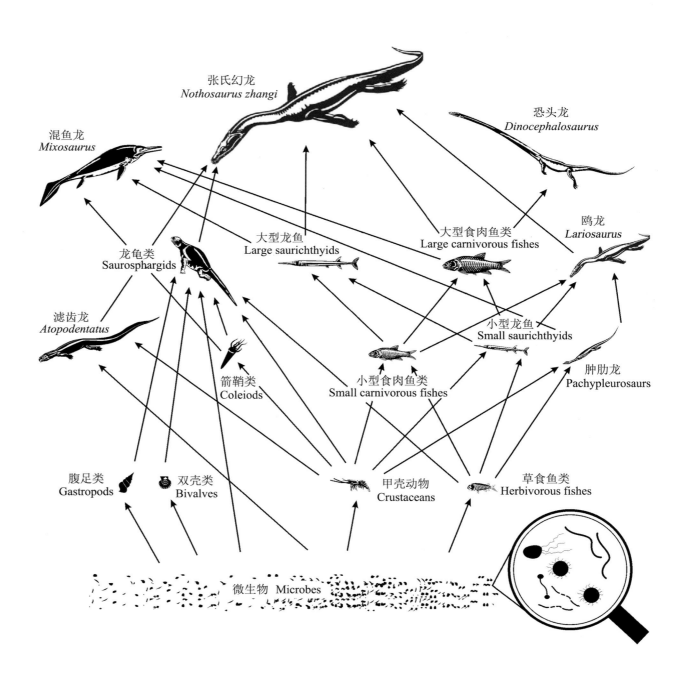

← 罗平生物群复原图, Brian Choo绘制
Reconstruction of the Luoping biota, by Brian Choo.

罗平生物群食物网复原图
Food web of the Luoping biota.

F 罗平生物群化石
ossils of the Luoping Biota

海生爬行类
Marine reptiles

罗平生物群目前发现的海生爬行类基本涵盖了三叠纪时期主要的海生爬行类别，包括鱼龙类、鳍龙类、原龙类以及初龙类。处于食物链顶层的海生爬行类的繁盛说明当时生态系统已经很完善，代表了二叠纪末生物大灭绝之后海洋生态系统的全面复苏。

Well-preserved, diverse marine reptiles are one of the highlights of the Luoping biota, including ichthyosaurs, sauropterygians, protorosaurs and archosauromorphs. The occurrence of a diverse assemblage of marine reptiles as top predators in the food web indicates a well developed marine ecosystem, showing full rebuilding of the marine ecosystem after the mass extinction that happened about 250 million years ago.

混鱼龙类
Mixosaurids

混鱼龙是一类繁盛于中三叠世的鱼龙。在罗平生物群的海生爬行类中，混鱼龙类的数量占统治地位。目前在罗平发现的鱼龙都属于中等大小的混鱼龙科。混鱼龙被认为是广泛觅食性动物，它们积极地四处巡游寻找自己的食物。罗平的混鱼龙至少有两种类型。其中一种是始祖异齿鱼龙，以同齿型为特征，主要捕食软躯体生物。另外一种和盘县混鱼龙极为相似，其异齿型指示这种鱼龙的食谱相对更为广泛，可能包括了软体动物、节肢动物以及鱼类。

Mixosaurids were a short-lived but extremely successful group of Middle Triassic ichthyosaurs. Mixosaurids dominates the Luoping marine reptiles in terms of individuals. Most mixosaurids from the Luoping biota are medium in size. As active predators, mixosaurids swam in the upper part of the water column and foraged for their prey. At least two types of mixosaurs are recongnized from the Luoping biota. The first type is *Phalarodon atavus*, characterized by homodont dentition with conical teeth, mainly used to feed on soft-bodied animals. The second type is similar to *Mixosaurus panxianensis*, characterized by heterodont dentition which indicates a wide spectrum of prey types, including molluscs, arthropods, and fishes.

盘县混鱼龙相似种，比例尺为5cm

Mixosaurus cf. *panxianensis*. Scale bar is 5cm.

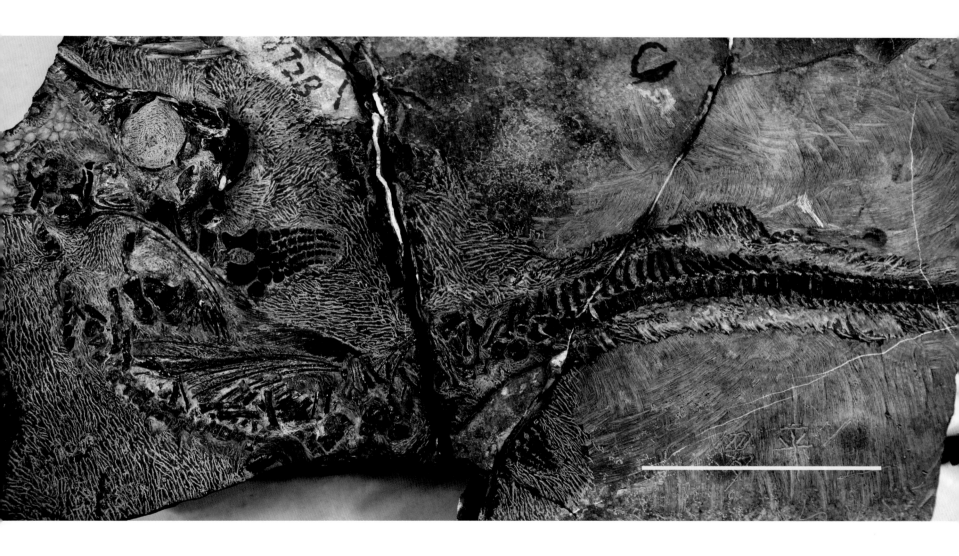

始祖异齿鱼龙，比例尺为10cm
Phalarodon atavus. Scale bar is 10cm.

原龙类
Protorosaurs

原龙类起源于古生代末期，繁盛于三叠纪中期，主要发现于古特提斯洋周缘的陆表海（现在的欧洲、中东及华南），以陆生类型为主，包括长颈龙、恐头龙、巨胫龙几个属。来自于罗平的原龙类为东方恐头龙，以加长的脖子为典型特征，是典型的水生动物，捕食鱼类等小型动物。

Protorosaurs originated in Late Palaeozoic and diversified in the Middle Triassic, and lived in shallow epicontinental seas along the margins of Pangea (modern Europe, the Middle East, and south China). Protorosaurs from the Luoping biota are identified as *Dinocephalosaurus orientalis*, a typical aquatic animal and characterized by its extremely long neck, which facilitated predation on fishes and other small prey.

东方恐头龙，比例尺为10cm
Dinocephalosaurus orientalis. Scale bar is 10cm.

初龙类
Archosaurs

罗平地区所见的初龙主要为一些散落的牙齿，这些牙齿的锯齿状边缘为初龙类的典型特征。在和罗平生物群时代相似的盘县动物群中产出大型的混形黔鳄。这种海生顶级捕食者具有匕首一样的牙齿，其体型指示其为典型的伏击型捕食者。这些特征使得它能够以极为迅猛的速度捕食一切靠近它的食物。罗平的这些初龙类的牙齿很有可能来自于黔鳄或者是亲缘关系非常接近的类群。

So far only a few disarticulated teeth of archosaurs have been recovered from the Luoping biota. The teeth show serrated margin characteristic of archosaurs. A complete and large archosaur, *Qianosuchus mixtus*, was reported from the coeval Panxian fauna in neighboring Guizhou Province and interpreted as a rapid predator because of its dagger-like teeth. It seems likely that the disarticulated archosaur teeth from Luoping were from *Qianosuchus* or a related animal.

初龙类牙齿，比例尺为1cm
A tooth of an archosaur. Scale bar is 1cm.

鳍龙类
Sauropterygians

罗平生物群鳍龙类包括小型的肿肋龙类、中等大小的鸥龙类以及巨型的幻龙类。另外两种和鳍龙类亲缘关系比较接近的来自罗平的海生爬行类是中等大小的龙龟类。其中一种叫云贵中国龙龟，外形酷似乌龟，因此得名龙龟。另外一种龙龟类被命名为大头龙。龙龟类可能为杂食性动物。

罗平生物群发现的一种奇特的海生爬行类奇特滤齿龙长有锤子头状的头，口腔中遍布细密的栅栏状牙齿，用于啃食和过滤藻类，是植食性海洋爬行动物的最早记录。其亲缘关系可能和鳍龙类比较接近。

最近发现的张氏幻龙是目前已知最大的三叠纪幻龙类，是罗平生物群中的巨型捕食者（指能够捕食与自身体型相当甚至更大的猎物）。巨型捕食者在罗平生物群的出现说明三叠纪中期的海洋生态系统已经全面复苏，并且是全球同步。

Sauropterygians from the Luoping biota include small-sized pachypleurosaurs, medium-sized *Lariosaurus*, and large-sized nothosaurids. Besides, two other types close to sauropterygians were also recovered from Luoping, including the turtle-like *Sinosaurosphargis yunguiensis* and *Largocephalosaurus polycarpon* with a relatively large skull. Unlike other predatory marine reptiles from Luoping, *Sinosaurosphargis* was probably an omnivore.

A special type of sauropterygian-related marine reptile, Atopodentatus unicus, was reported from Luoping. It has a hammer-like head, and two types of teeth: the chisel-shaped teeth and the densely packed needle-shaped teeth. The chisel-shaped teeth were used to scrape algae off the substrate, and the plant matter was sucked in and filtered by the closely packed needle-shaped teeth. This is the oldest record of herbivory within marine reptiles.

A recently reported new nothosaurid, *Nothosaurus zhangi*, is the largest known Triassic sauropterygians and was interpreted as macropredator in the food web. Macropredators are capable of hunting, seizing, and dismembering prey of equal or even larger body size than their own. The occurrence of macropredators in the Luoping biota indicates globally synchronous recovery of shallow marine ecosystems by the time of Middle Triassic.

丁氏滇龙模式标本，比例尺为1cm
Holotype of *Dianopachysaurus dingi*. Scale bar is 1cm.

丁氏滇龙，比例尺为1cm

Dianopachysaurus dingi. Scale bar is 1cm.

利齿滇东龙，硬币为比例尺
Diandongosaurus acutidentatus Coin for scale.

鸥龙类未定种，比例尺为10cm
Lariosaurus sp. Scale bar is 10cm.

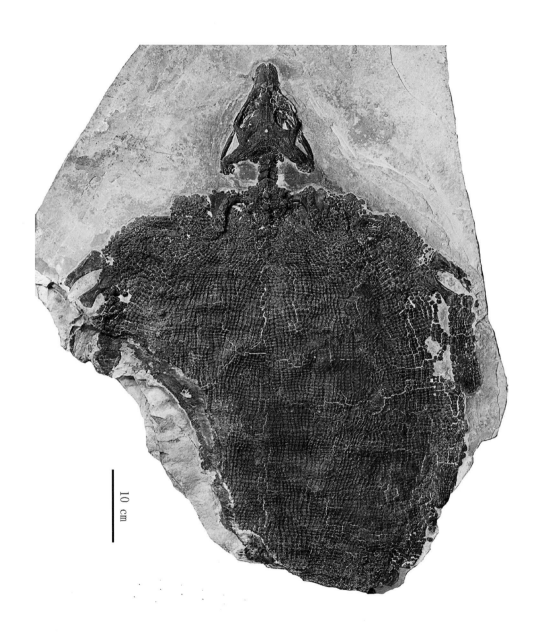

云贵中华龟龙，比例尺为10cm
Sinosaurosphargis yunguiensis. Scale bar is 10cm.

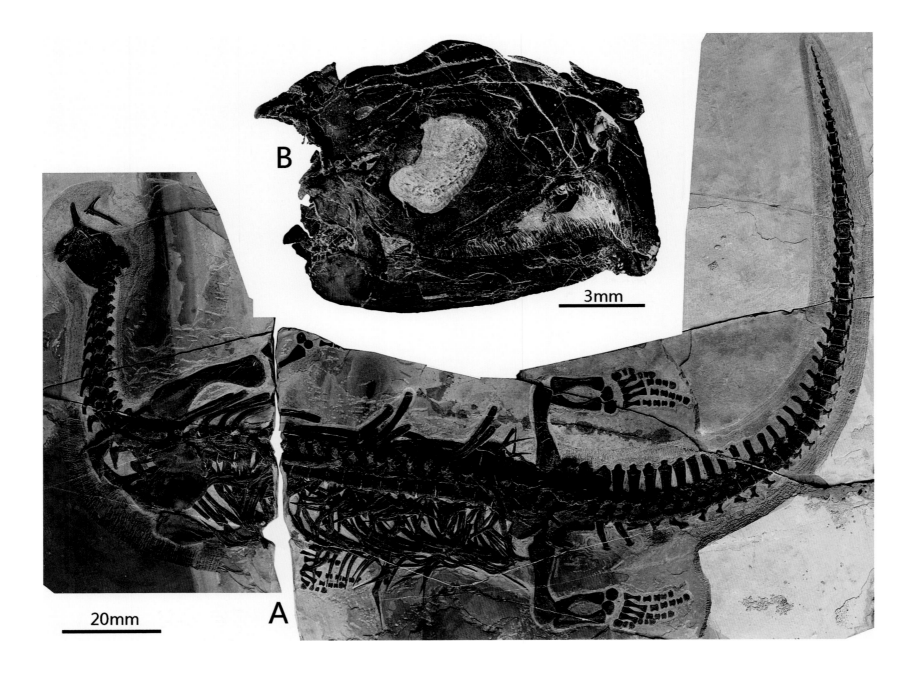

奇特滤齿龙。A. 骨骼全貌；B. 头骨反面放大，图片由程龙提供
Atopodentatus unicus. A.Overview of the whole body; B.Enlargement of skull from underlying side. Photo courtesy of Long Cheng.

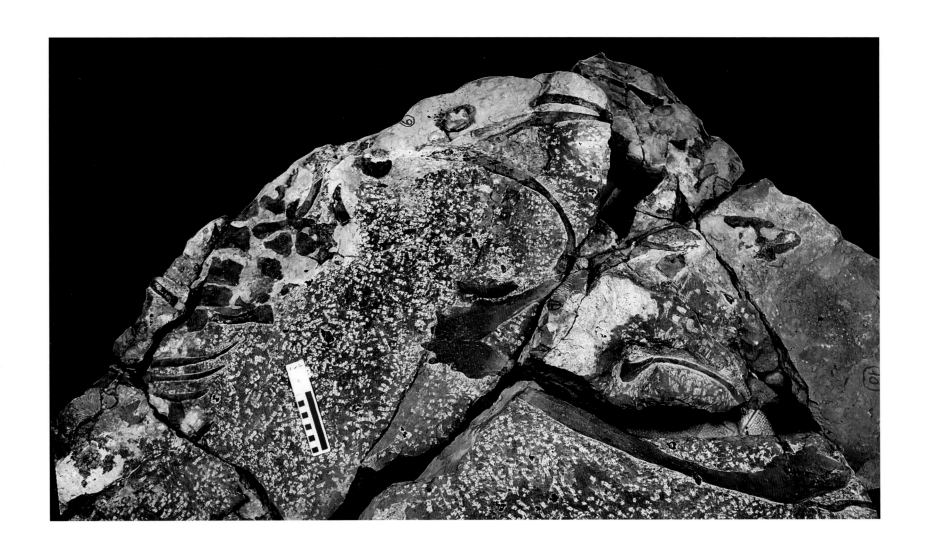

张氏幻龙下颚，已知最大的三叠纪幻龙下颚，比例尺为10cm
Nothosaurus zhangi, the largest known lower jaw among Triassic sauropterygians. Scale bar is 10cm.

张氏幻龙捕食情景复原图，Brian Choo 绘制

Reconstruction of predation by *Nothosaurus zhangi*, by Brian Choo.

鱼类
Fishes

鱼类是罗平生物群最具特色的动物化石类型，数量丰富，种类繁多，包括至少30余种，其中大部分为新属新种。目前已经命名的辐鳍鱼类化石已有19种，分别属于辐鳍鱼亚纲软骨硬鳞鱼次纲的古鳕鱼类、龙鱼类、裂齿鱼类、肋鳞鱼类，辐鳍鱼亚纲新鳍次纲的半椎鱼类和鲱口类等。肉鳍鱼类有2个种。罗平生物群新鳍鱼类数量多，种类丰富，证实该类群在安尼期就已经开始了辐射。

The abundant fishes with high diversity are an important component of the fauna, including more than 30 taxa. Initial identifications indicate more than 9 families of actinopterygian fishes, and 19 new species belonging to Palaeoniscidae, Saurichthyidae, Peltopleuridae, Habroichthyidae, Perleididae, Polzbergidae, Semionotidae, Marcopoloichthyidae, Halecomorphi and Ginglymodi have been erected. There are two species of Coelacanthiformes. Many of the remaining Luoping fishes are undescribed taxa, which are currently under study. The occurrence of diverse and abundant neopterygians (meaning 'new' fishes) in the Luoping biota suggests that a first major radiation of neopterygian fishes already took place during the early Middle Triassic, much earlier than had been thought until recently.

亚全骨鱼类
The "Subholosteans"

二叠纪末生物大灭绝后，由于早三叠世的极端环境，鱼类直到中三叠世安尼期才开始辐射。中国下扬子地区的三叠纪地层拥有了相对完整的鱼化石记录。安徽巢湖、句容等地早三叠世鱼类化石相对单一，而且地域性很强。但是安尼期罗平生物群的鱼类化石种类丰富，保存精美，主要由原始辐鳍鱼类、亚全骨鱼类、新鳍鱼类基干类群和空棘鱼类组成。亚全骨鱼类是古鳕鱼类和新鳍鱼类之间的过渡类型，同时拥有了有原始和进步的特征。罗平生物群亚全骨鱼类主要由裂齿鱼目和肋鳞鱼目组成。

After the end-Permian Mass Extinction, the radiation of marine fishes didn't happen until the early Anisian because of the extreme environment in the early Triassic. The Lower Yangtze region witnesses a relatively continuous sequence of fish records. Fish fossils from the early Triassic assemblage in Chaohu, Jurong are less diverse and endemic. However, the Anisian Luoping biota bears various fish fossils with high-quality preservation, consisting of basal actinopterygians, subholosteans, basal neopterygians as well as coelacanths. Subholosteans, as the transitional forms between Paleoniscoid fishes and Neopterygii ones, always have both primitive and advanced characters. The subholosteans in the Luoping biota mainly includes Perleidiformes and Peltopleuriformes.

裂齿鱼目
Perleidiformes

裂齿鱼目定名的有5属，分别是罗平鱼属、罗平裂齿鱼属、裂齿鱼属、富源裂齿鱼属和滇东裂齿鱼属。

Five genera of Perleidiformes have been published from Luoping: *Luopingichthys*, *Luopingperleidus*, *Perleidus*, *Fuyuanperleidus* and *Diandongperleidus*.

贝氏罗平鱼,比例尺为1cm
Luopingichthys bergi. Scale bar is 1cm.

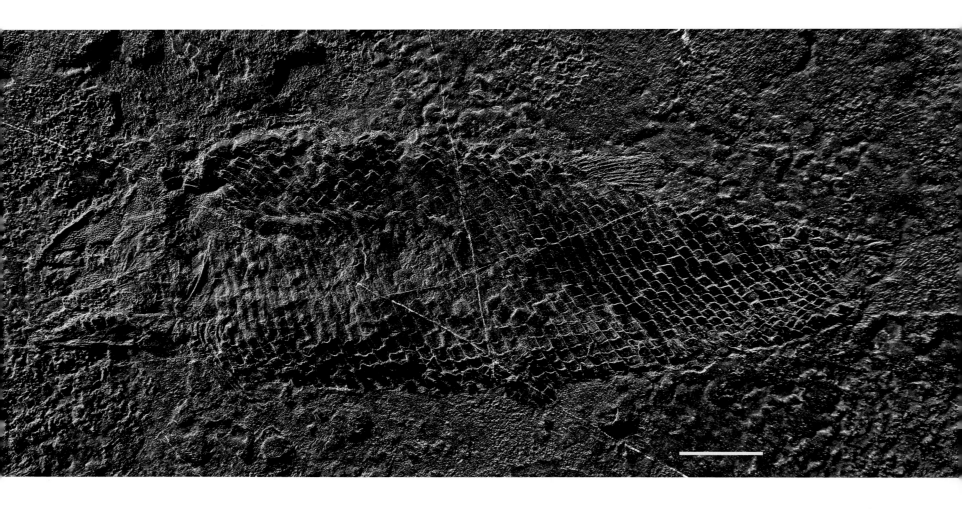

小齿滇东裂齿鱼，比例尺为2cm
Diandongperleidus denticulatus. Scale bar is 2cm.

中华裂齿鱼，比例尺为1cm
Perleidus sinensis. Scale bar is 1cm.

苏氏罗平裂齿鱼,比例尺为1cm
Loupingperleidus sui. Scale bar is 1cm.

邓氏富源裂齿鱼，比例尺为1cm
Fuyuanperleidus dengi. Scale bar is 1cm.

肋鳞鱼目
Peltopleuriformes

　　肋鳞鱼目属于软骨硬鳞次纲，也是"亚全骨鱼类"中的典型类群。从中三叠世才开始出现，部分属种可延续到早侏罗世。肋鳞鱼类在云南和贵州中三叠统地层中很常见，而且数量繁多。肋鳞鱼在罗平生物群鱼类化石中占了很大的比例，是最为普遍的软骨硬鳞鱼类之一，大部分为*Habroichthys*与*Placopleurus*。

Peltopleuriformes in Luoping mainly consist of *Habroichthys*, *Placopleurus* and *Peripeltopleurus*. *Habroichthys* and *Placopleurus* were the commonest subholoteans in the Luoping biota. There are two species belonging to *Habroichthys*, *Habroichthys broughi* and *Habroichthys minimus*. These little fish range widely from Middle Anisian to Late Ladinian in South China.

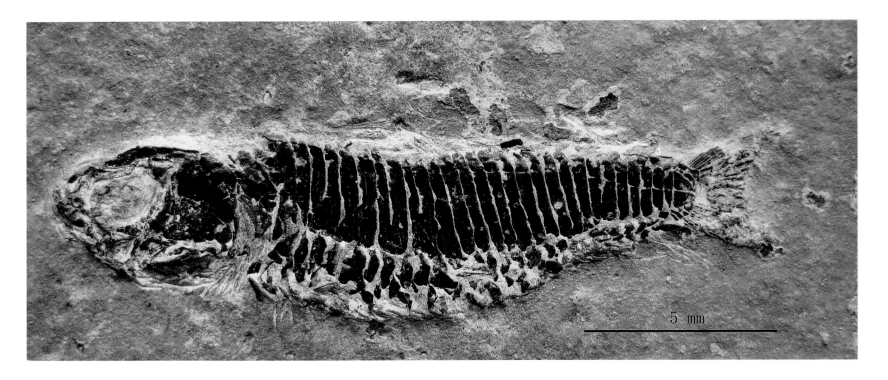

包氏海博鱼
Habroichthys broughi.

包氏海博鱼，比例尺为1cm
Habroichthys broughi. Scale bar is 1cm.

海博鱼复原图，陈庆韬绘制
Reconstruction of *Habroichthys broughi*, by Qingtao Chen.

海博鱼未定种，比例尺为1cm
Habroichthys sp. Scale bar is 1cm.

肋鳞鱼未定种，比例尺为1cm
Peripeltopleurus sp. Scale bar is 1cm.

软骨硬鳞鱼类
Chondrostei

软骨硬鳞类是辐鳍鱼类的基干类群，具有较原始的特征，例如上颌骨与前鳃盖骨，几乎不能活动，间鳃盖不存在，鳍条多于支持骨，尾鳍为歪型或半歪型。罗平生物群的软骨硬鳞鱼类包括古鳕目和龙鱼目。

Chondrostei is the basal group of actinopterygians. They had many primitive characters: maxilla fixed to the preopercular, interopercular absent, unpaired fin rays more than their supports, heterocercal or hemiheterocercal tail. Chondrostei of the Luoping Biota consist of Palaeonisciformes and Saurichthyiforms.

古鳕目
Palaeonisciformes

古鳕目属于辐鳍亚纲中软骨硬鳞次纲，是辐鳍亚纲中的基干类群，也是原始辐鳍鱼类中最具代表性的繁盛类群，其化石记录最早见于中泥盆世地层，石炭纪和二叠纪最为繁盛，到早白垩世灭绝。

尼尔森翼鳕代表了翼鳕属在亚洲的首次发现，它的发现为支持三叠纪时期东、西特提斯洋之间存在生物交流的假说提供了新证据。另外，作为翼鳕属最晚的代表种之一，尼尔森翼鳕的发现表明翼鳕属在早三叠世末并没有灭绝，至少延续到中三叠世早期。

As the stem group of Actinopterygii, palaeonisciforms are classified in Chondrostei. The earliest fossil palaeonisciforms are found from rocks of the Middle Devonian. They diversified in the Carboniferous and Permian, decreased thereafter and went extinct in the Early Cretaceous.

A new species is the first record of *Pteronisculus* in Asia, providing new evidence to support the Triassic biological exchanges between the eastern and western Palaeotethys Ocean. Moreover, as one of the youngest members of *Pteronisculus*, the new finding provides convincing evidence to show that *Pteronisculus* lived through the Early Triassic and survived at least at the early stage of the Middle Triassic.

尼尔森翼鳕模式标本，比例尺为1cm，图片由徐光辉提供
Holotype of *Pteronisculus nielseni*. Scale bar is 1cm. Photo courtesy of Guanghui Xu.

尼尔森翼鳕，比例尺为1cm，图片由徐光辉提供
Pteronisculus nielseni. Scale bar is 1cm. Photo courtesy of Guanhui Xu.

龙鱼目
Saurichthyiformes

龙鱼目属于辐鳍亚纲(Actinopterygii)中的软骨硬鳞次纲（Chondrostei）。为辐鳍亚纲的基干类群。龙鱼体长，吻极突出，牙齿强大，身披4列或6列鳞，上颌形状与古鳕类相似，鳃盖部由单一的大的鳃盖骨构成。龙鱼拥有流线型的身体，游泳能力极强。在罗平生物群中产出的大量的龙鱼类化石，保存完好，个体数量极其丰富。而且其中包含了从幼年个体到成年个体及其过渡类型。最小的为2cm，最长的达72cm。罗平生物群龙鱼类化石高度分异，目前命名有2属4种。到目前为止，罗平生物群中龙鱼类化石已鉴定发表的有大洼子龙鱼（*Saurichthys dawaziensis*）、云南龙鱼（*Saurichthys yunnanensis*）、长奇鳍中华龙鱼（*Sinosaurichthys longimedialis*）和小型中华龙鱼（*Sinosaurichthys minuta*）。

Saurichthyiformes are a stem group of Actinopterygii. They have an elongated body, a prominent jaw with strong dentition, nearly naked body with four to six scale rows, the shape of maxilla is palaeoniscid-like, and there is only one big opercular. The streamlined body makes them fast swimmers. Saurichthyiformes in the Luoping biota are abundant and well preserved. The smallest juvenile is only 2cm in length, whereas the largest adult is 72cm in length. So far, 4 species of 2 genera have been described, namely *Saurichthys dawaziensis*, *Saurichthys yunnanensis*, *Sinosaurichthys longimedialis*, and *Sinosaurichthys minuta*.

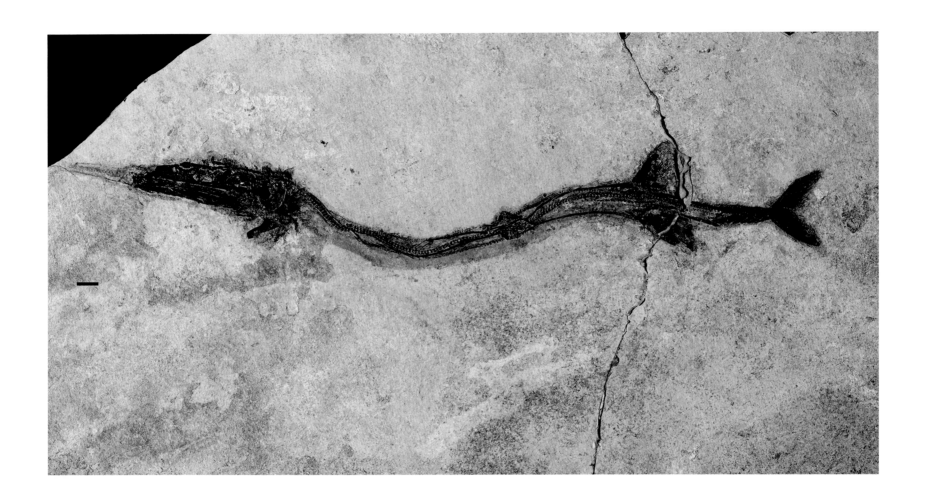

大洼子龙鱼，比例尺为1cm
Saurichthys dawaziensis. Scale bar is 1cm.

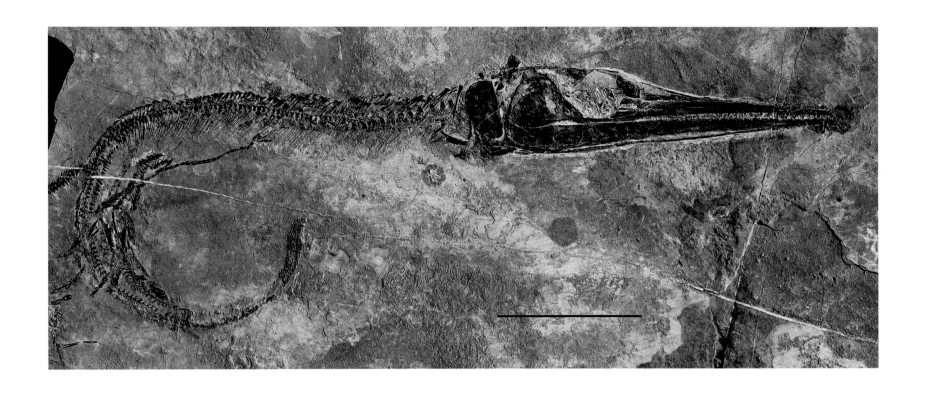

云南龙鱼，比例尺为5cm
Saurichthys yunnanensis. Scale bar is 5cm.

长奇鳍中华龙鱼，比例尺为5cm
Sinosaurichthys longipectoralis. Scale bar is 5cm.

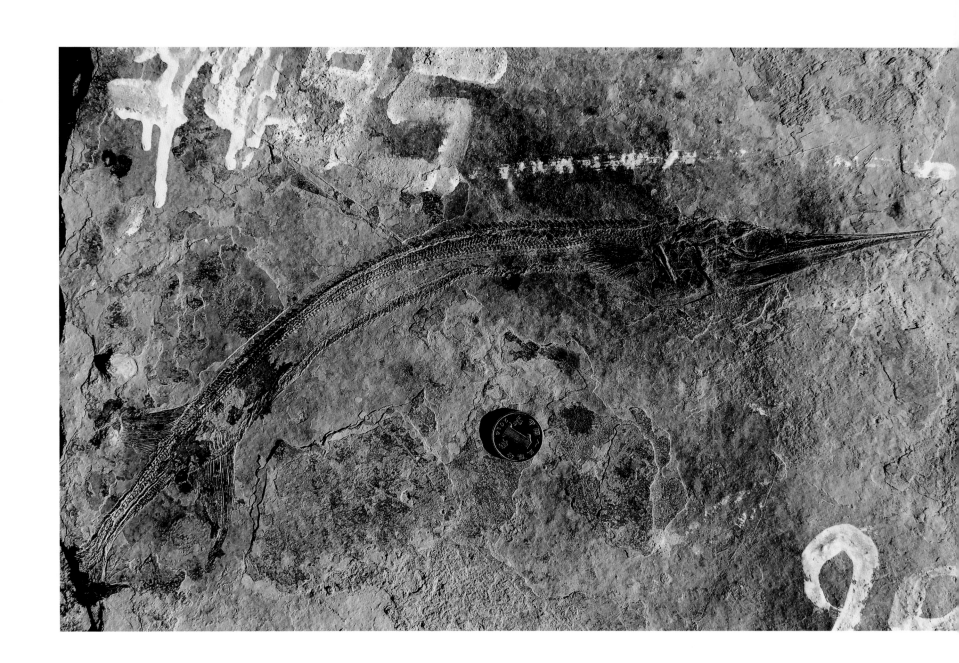

长奇鳍中华龙鱼，硬币为比例尺
Sinosaurichthys longipectoralis. Coin for scale.

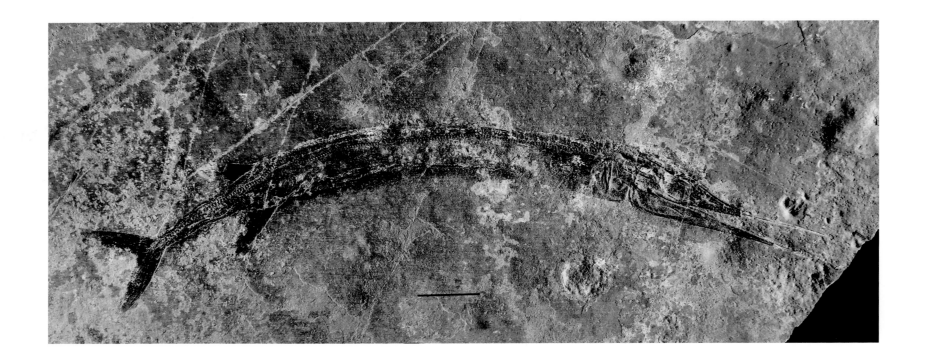

小型中华龙鱼，比例尺为2cm
Sinosaurichthys minuta. Scale bar is 2cm.

小型中华龙鱼，比例尺为2cm
Sinosaurichthys minuta. Scale bar is 2cm.

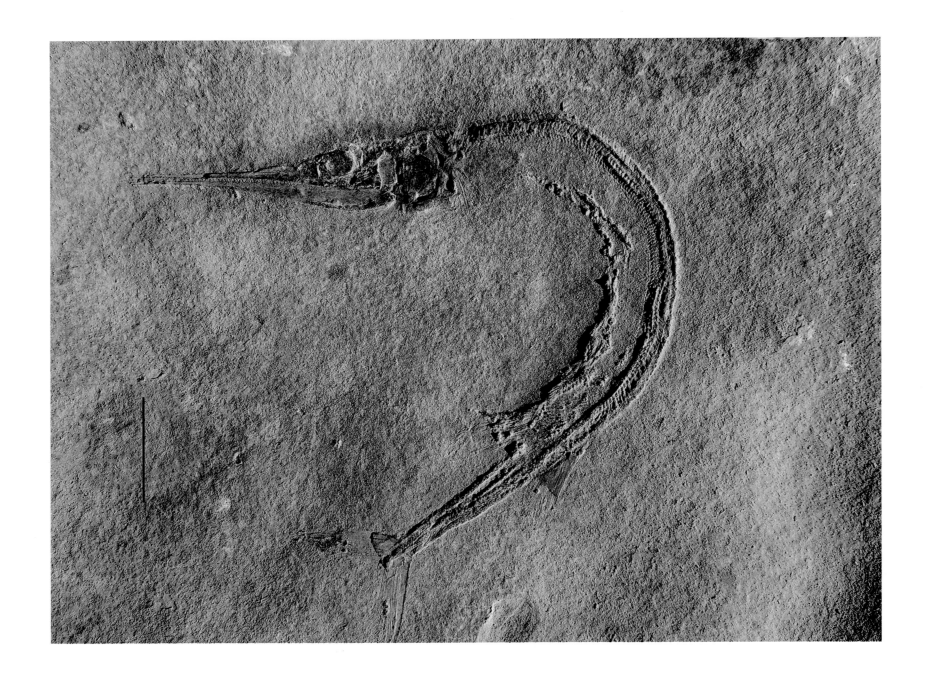

小型中华龙鱼，比例尺为2cm
Sinosaurichthys minuta. Scale bar is 2cm.

小型中华龙鱼，比例尺为2cm
Sinosaurichthys minuta. Scale bar is 2cm.

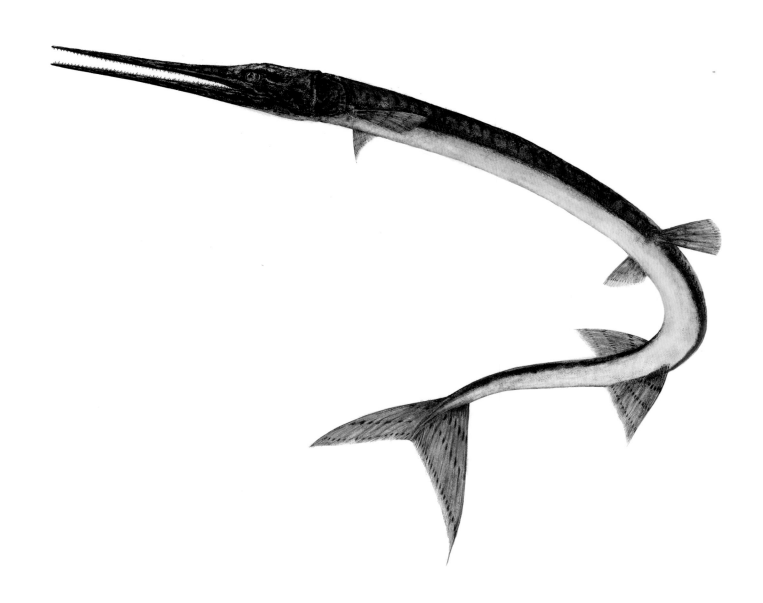

小型中华龙鱼复原图，陈庆韬绘制
Reconstruction of *Sinosaurichthys minuta*, by Qingtao Chen.

新鳍鱼类
Neopterygians

辐鳍鱼类进化中非常重要的一步就是新鳍鱼类的出现。新鳍鱼类出现在晚二叠时期，繁盛于中生代，一直延续到现代的真骨鱼类。新鳍鱼类与原始辐鳍鱼类在结构上有了很大的进步。上颌骨不再与前鳃盖相铰接，下颌具有冠状突，前鳃盖几乎垂直且背枝细长，具间鳃盖，每块支持骨只支持一根鳍条，尾鳍半歪尾型。支持骨与鳍条对应关系的改变使得新鳍鱼类能更好地控制背鳍和臀鳍，有了更灵活的游泳能力。颌部结构的改变使它们有强的咬合力，捕食能力得到加强。

罗平生物群中新鳍鱼类包含有体型异化的高背罗雄鱼和格兰德拱鱼，有身体几乎无鳞的意外裸鱼和安氏马可波罗鱼，最普遍的苏氏圣乔治鱼，还有罗平强壮鱼。罗雄鱼与拱鱼有着相似的身体轮廓和相似的头部结构。这类高纺锤体型的鱼类在游泳时受到的阻力较大，所以游泳速度并不快。安氏马可波罗鱼身体几乎没有鳞片，是一类非常特殊的新鳍鱼类。它的颌部类似更为先进的真骨鱼类，前鳃盖呈"L"形，而且上鳍条的支持骨上也存在异化。让人惊奇的是这类小型鱼类居然有着非常广泛的分布范围，在古特提斯洋的东西两岸都有分布。罗平强壮鱼的发现代表了迄今为止最早的预言鱼目化石记录，把预言鱼目的时间往前延伸到了中三叠世，比原来的记录往前提前了9000万年，同时扩大了它的地理分布范围，从欧洲到古特提斯海洋东岸的中国南方。苏氏圣乔治鱼以独特的围眶骨为特点，它是罗平生物群中数量最多的新鳍鱼类，在鱼化石富集的层面上每平方米平均有20多条。多饰维纳斯鱼代表了新鳍鱼类第二性征的最早化石记录，对研究新鳍鱼类的行为、繁殖方式和早期分异提供了重要信息。

The presence of neopterygians is the most important step during the evolution of actinopterygians. They first appeared in the Late Permian, peaked in the Mesozoic, and continue as modern Holostei. Neopterygians are more derived than the primitive actinopterygians, in the following aspects: maxilla no longer hinged with the preopercular, mandible has coronoid process, preopercular almost vertical with slender dorsal branch, interopercular present, unpaired fin rays about equal in number to their supports, hemiheterocercal tail. These changes enhance their ability in swimming and preying.

The neopterygians of the Luoping biota include deep-bodied *Luoxiongichthys hyperdorsalis*, *Kyphosichthys grandei* and naked *Gymnoichthys inoponatus*, *Marcopoloichthys ani*, abundant *Sangiorgioichthys sui*, as well as *Robustichthys luopingensis*. *Luoxiongichthys* and *Kyphosichthys* have similar deep bodies and have many similarities in their skulls. These deep-bodied fishes have a large surface that causes a large drag during swimming. So, they are not good swimmers. *Marcopoloichthys ani* is a unique naked small neopterygian. It has a modified teleost-like jaw system, "L"-shaped preopercle and modified support of the medial fin. It is surprising that this small-sized fish shows a very wide distribution, found both in eastern and western Paleo-Tethys. *Robustichthys luopingensis* documents the oldest known ionoscopiform, extending the stratigraphic range of this group by approximately 90 million years, and the geographical distribution of this group into the Middle Triassic of South China, a part of the eastern Palaeotethys Ocean. *Sangiorgioichthys sui* has unique circumorbital bones. It is the most abundant neopterygian in the Luoping biota, with more than 20 specimens of *Sangiorgichthys* on average per square metre on certain layers of Dawazi quarry. The finding of sexual dimorphism in *Venusichthys comptus* provides an important addition for understanding the behavior, reproduction, and early diversification of Neopterygii.

高背罗雄鱼模式标本，比例尺为1cm
Holotype of *Luoxiongichthys hyperdorsalis*. Scale bar is 1cm.

高背罗雄鱼复原图，Brian Choo绘制
Reconstruction of *Luoxiongichthy hyperdorsalis*, by Brian Choo.

高背罗雄鱼，比例尺为2cm
Luoxiongichthys hyperdorsalis. Scale bar is 2cm.

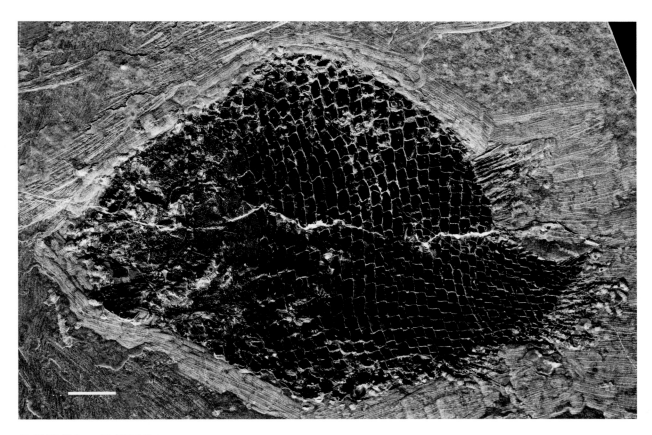

格兰德拱鱼，比例尺为1cm
Kyphosichthys grandei. Scale bar is 1cm.

格兰德拱鱼复原图，Brian Choo绘制
Reconstruction of *Kyphosichthys grandei*, by Brian Choo.

意外裸鱼，比例尺为1cm
Gymnoichthys inoponatus. Scale bar is 1cm.

意外裸鱼复原图，Brian Choo绘制
Reconstruction of *Gymnoichthys inoponatus*, by Brian Choo.

安氏马可波罗鱼，比例尺为1cm
Marcopoloichthys ani. Scale bar is 1cm.

安氏马可波罗鱼，比例尺为1cm
Marcopoloichthys ani. Scale bar is 1cm.

罗平强壮鱼，比例尺为1cm
Robustichthys luopingensis. Scale bar is 1cm.

苏氏圣乔治鱼，比例尺为1cm
Sangiorgioichthys sui. Scale bar is 1cm.

苏氏圣乔治鱼，比例尺为1cm
Sangiorgioichthys sui. Scale bar is 1cm.

苏氏圣乔治鱼,比例尺为1cm
Sangiorgioichthys sui. Scale bar is 1cm.

苏氏圣乔治鱼，比例尺为1cm
Sangiorgioichthys sui. Scale bar is 1cm.

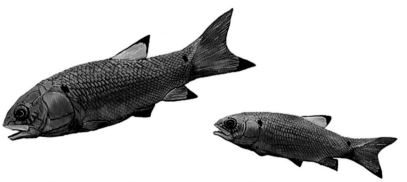

苏氏圣乔治鱼复原图，Brian Choo绘制
Reconstruction of *Sangiorgioichthys sui*, by Brian Choo.

多饰维纳斯鱼雄性个体，比例尺为1cm
Venusichthys comptus, male. Scale bar is 1cm.

多饰维纳斯鱼雄性个体,图片由徐光辉提供
Venusichthys comptus, male. Photo courtesy of Guanghui Xu.

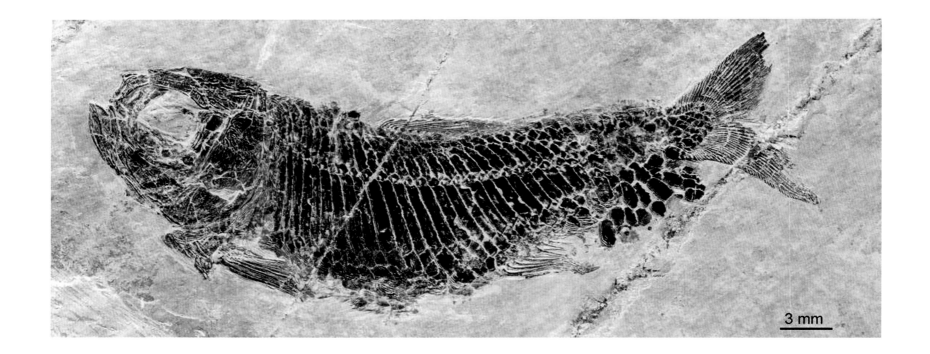

多饰维纳斯鱼模式标本，雌性个体，图片由徐光辉提供
Holotype of *Venusichthys comptus*, female. Photo courtesy of Guanghui Xu.

空棘鱼类
Coelacanths

空棘鱼属于硬骨鱼纲肉鳍亚纲，繁盛于三叠纪，具两个背鳍和一对较大的喉板骨，尾鳍分三页。鳞片呈舌形，覆盖部分有同心纹，露出部分后缘有脊状纹饰。曾经人们认为它已经灭绝，直到1938年才在非洲南部东海岸发现了第一条现生的拉蒂迈鱼。在经历了两亿多年的时间，几乎没有进化，由此称它为"活化石"。

罗平生物群中空棘鱼类相对稀少，现已发现两个新属新种，即宽泪颧骨罗平空棘鱼和尖瘤云南空棘鱼。宽泪颧骨罗平空棘鱼标本内的两个胚胎显示了该类动物最早的卵胎生证据，把晚侏罗世的记录提前到了中三叠世。

Coelacanthiformes are bony fishes, members of the lobe-finned Sarcopterygii. Coelacanths were diverse in the Triassic. They possess two dorsal fins, big paired gulars and a three-lobed caudal fin. Ligulate scales decorated by fine, parallel striae on the covered region and hollow ridges in the exposed region. They were thought to be extinct until 1938, when a modern *Latimeria* was found off the east coast of South Africa. Traditionally, the coelacanth was considered a "living fossil" due to its apparent lack of significant evolution over the past millions of years.

Coelacanths are rare in the Luoping biota. Two of them have been named, *Luopingcoelacanthus eurylacrimalis* and *Yunnancoelacanthus acrotuberculatus*. The two embryos found in *Luopingcoelacanthus eurylacrimalis* is the oldest record of ovoviviparity in coelacanths, extending the evidence from the Late Jurassic to the Middle Triassic.

尖瘤云南空棘鱼，比例尺为2cm

Yunnancoelacanthus acrotuberculatus. Scale bar is 2cm.

宽泪颧骨罗平空棘鱼，比例尺为5cm
Luopingcoelacanthus eurylacrimalis. Scale bar is 5cm.

空棘鱼复原图，Brian Choo绘制
Reconstruction of *Luopingcoelacanthus eurylacrimalis*, by Brian Choo.

节肢动物
Arthropods

　　节肢动物是罗平生物群数量最为丰富的门类，目前发现的化石以甲壳亚门为主，另外还发现有螯肢亚门的鲎类化石和多足亚门的千足虫类化石。鲎、千足虫、等足类等化石都是首次在我国发现。

Arthropods dominate the Luoping biota in terms of individuals, including crustaceans, millipedes and horseshoe crabs. Fossil horseshoe crabs, millipedes, and isopods are recovered from China for the first time.

甲壳亚门
Crustacea

甲壳类节肢动物是罗平生物群中数量最为丰富的化石类群，占整个生物群个体数量的90%以上。包括了十足目、等足目、糠虾类、叶肢介以及可能属于甲壳纲的原蟹类等。十足目的种类繁多，目前已发现约10余个物种，绝大部分为新属新种。数量上则以个体较小的糠虾最为丰富，是当时海洋生态系统食物链底层的主要分子，为处于食物链中部的食肉鱼类提供了充足的食物来源，这些鱼类又成为海生爬行类等顶级捕食者的食物。罗平生物群中大量保存完好、种类多样的甲壳类节肢动物说明该类动物在三叠纪中期开始了快速的演化辐射。其中的十足目龙虾类化石是目前已知保存最为完好、数量最丰富、时代最早的龙虾化石群之一。

Crustaceans are the most common and diverse arthropod group of the Luoping biota, represented by decapods, isopods, cycloids, mysidiaceans, conchostracans and ostracods, amounting to 90% of the biota in terms of individuals. At least 10 species of decapods have been recognized. Abundant mysidiaceans form the basis of the food web. They were preyed on by carnivorous fishes, which in turn were fed upon by top predators, mostly marine reptiles. The occurrence of diverse and abundant crustaceans in the Luoping biota indicates a Middle Triassic radiation of this group. It also worthy of note that the Luoping lobsters represent one of the earliest and most diverse fossil lobster assemblages known so far.

十足目
Decapoda

十足目是罗平生物群节肢动物中最引人注目的类群，包括虾类和龙虾两大类群。十足目节肢动物化石主要发现于中生代和新生代，古生代仅有3个种被确认。罗平生物群中的十足目化石数量丰富、种类多样且保存精美，目前已经报道的有3个科的龙虾和2个科的虾类，包括至少5个属7个种，代表了十足目节肢动物在三叠纪中期的辐射演化，具有重要的研究意义。

Decapods are one of the most attractive groups of the Luoping biota, including lobsters and shrimps. Fossil decapods are mainly found from Mesozoic and Cenozoic rocks, whereas only 3 putative species have been reported from the Paleozoic. Fossil decapods from the Luoping biota are well preserved, abundant and diverse. So far, 3 lobster families and 2 shrimp families have been identified from the Luoping biota, including at least 5 genera and 7 species, representing the radiation of decapod arthropods since at least the Middle Triassic.

罗平棒手龙虾，比例尺为1cm
Koryncheiros luopingensis. Scale bar is 1cm.

中华三指龙虾,比例尺为1cm
Tridactylastacus sinensis. Scale bar is 1cm.

中华三指龙虾,比例尺为1cm
Tridactylastacus sinensis. Scale bar is 1cm.

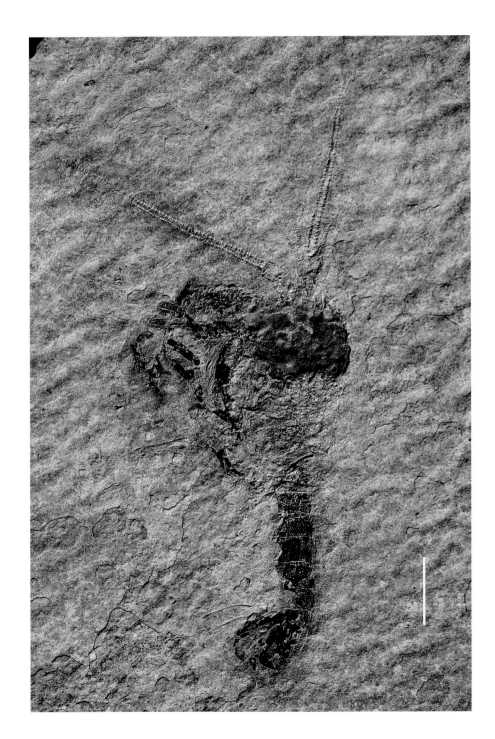

施氏云南龙虾，比例尺为1cm
Yunnanopalinura schrami. Scale bar is 1cm.

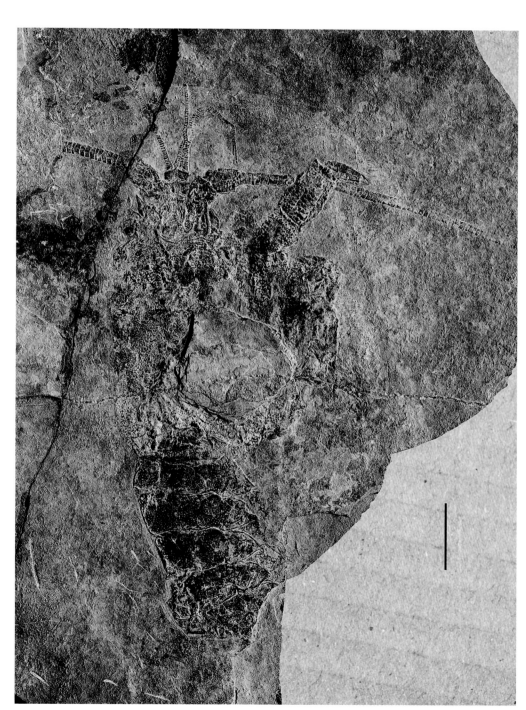

施氏云南龙虾，比例尺为1cm
Yunnanopalinura schrami. Scale bar is 1cm.

短额安尼伊吉虾，比例尺为1cm
Anisaeger brevispinus. Scale bar is 1cm.

具刺安尼伊吉虾，比例尺为5mm
Anisaeger spiniferus. Scale bar is 5mm.

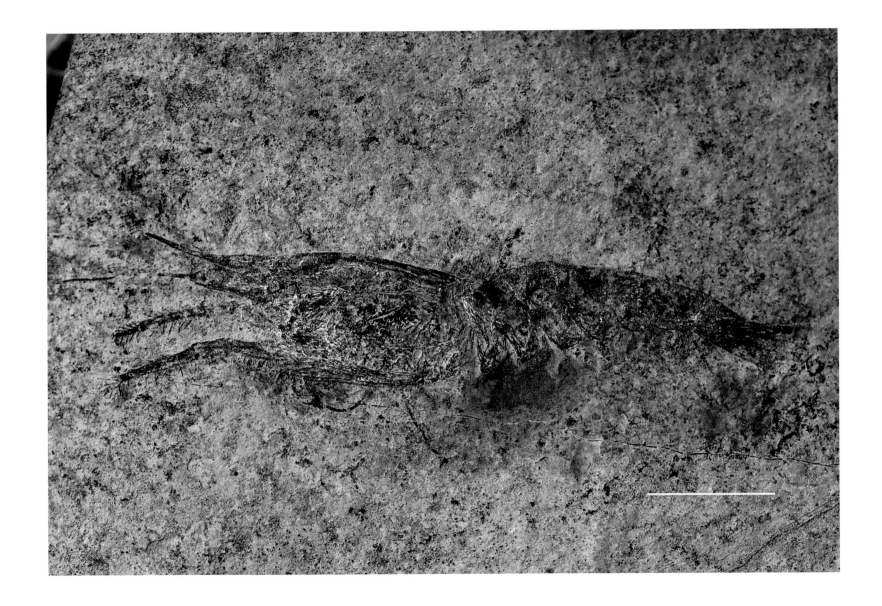

泸西伊吉虾，比例尺为1cm

Aeger luxii. Scale bar is 1cm.

壮美异形伊吉虾，比例尺为2cm
Distaeger prodigiosus. Scale bar is 2cm.

壮美异形伊吉虾，比例尺为1cm
Distaeger prodigiosus. Scale bar is 1cm.

糠虾类
Mysidaceans

糠虾属于节肢动物门软甲亚纲，包括3个目，糠虾类最早的化石记录来自中国和法国的三叠纪地层。糠虾是罗平生物群中数量最为丰富的化石，形成了当时生态系统的基石，为其他动物提供了充足的食物来源。罗平及三叠纪其他的糠虾从形态上看与现生的种类十分类似，说明现生糠虾类的躯体构型至少在三叠纪时期就已经实现。

Mysidaceans are a group of shrimp-like malacostracan crustaceans, including three orders, Mysida, Lophogastrida, and Stygiomysida. The earliest record of the Mysidacea is in the Triassic of France and China. Mysidaceans are the most abundant animals of the Luoping biota in terms of individuals, forming the base of the food web and are food resources for animals at the higher levels. The Luoping and other Triassic mysidaceans demonstrate remarkable similarities to the mysids of today, suggesting that the extant mysidaceans had achieved stable body plans relatively early in their history.

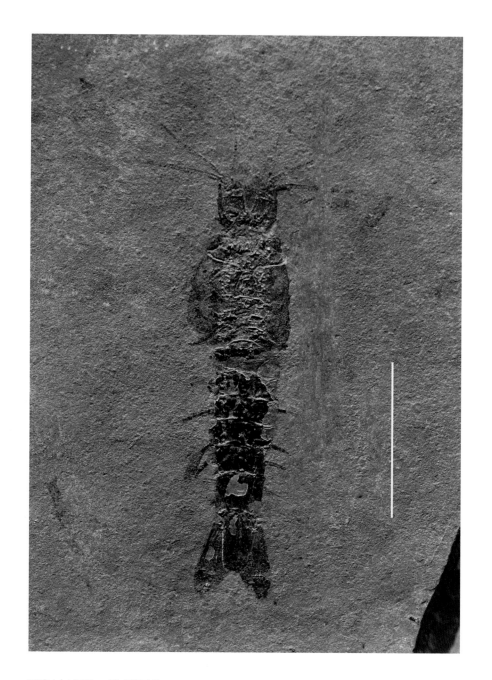

糠虾未定种，比例尺为1cm
An unnamed mysids. Scale bar is 1cm.

糠虾未定种，比例尺为1cm
An unnamed mysids. Scale bar is 1cm.

等足目
Isopods

等足目化石也是罗平生物群节肢动物的特色之一。此前等足目化石主要发现于欧洲、北美和澳大利亚。罗平生物群是目前该类化石在我国的唯一已知产地。罗平生物群等足目化石曾经被认为是淡水种类,是由河流带入海中的。但从其与大量典型海生动物化石共生且缺少搬运证据来看,罗平动物群内的等足类应该属于海生类型。目前罗平生物群已经报道的等足目化石有一个种,即白氏原双节虫。

Fossil isopods were only reported from Europe, North America and Australia before the discovery of the Luoping biota. Isopods from the Luoping biota are interpreted as terrestrial and transported into the sea by river currents. However, their co-occurrence with other typical marine animals, and the absence of transportation evidence, indicates that the Luoping isopods might have been marine rather than fresh water species. So far only a single taxon, *Protamphisopus baii*, has been described.

白氏原双节虫，比例尺为5mm
Fossil isopod, *Protamphisopus baii*. Scale bar is 5mm.

白氏原双节虫，比例尺为1cm
Fossil isopod, *Protamphisopus baii*. Scale bar is 1cm.

囊头类
Thylacocephalans

囊头类是一类游泳的肉食性节肢动物，可能和甲壳类有亲缘关系。该类化石最典型的特征包括分节的壳、一对巨大的眼睛，以及数对爪状的长附肢。其最早的化石记录出现在寒武纪，三叠纪时期在特提斯大洋东西两侧均有发现，该类动物在当时的生态系统中起着重要作用。

Thylacocephalans are pelagic and predatory arthropods, with a possible phylogenetic relationships with crustaceans. The most diagnostic characteristics include the segmented carapace, a pair of large eyes and several pairs of long raptorial appendages. The fossil record of thylacocephalans can be dated back to Early Cambrian. The worldwide distribution of thylacocephalans during Triassic suggest a important role in marine communities.

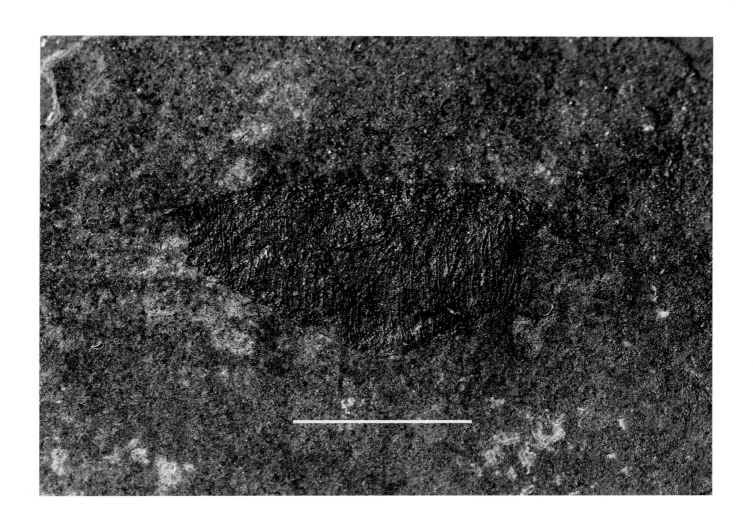

囊头类，比例尺为1cm
An unnamed thylacocephalans. Scale bar is 1cm.

原蟹类
Cycloids

原蟹类是地质历史上出现过的一类非常奇特的节肢动物，它们的化石发现在晚古生代至中生代的海相地层中。虽然名为原蟹类，但是原蟹类与现生螃蟹之间没有直接的亲缘关系，它们仅仅外形上类似。有趣的是，在真正的螃蟹出现之前（前侏罗纪）的生态系统中，原蟹类起到了类似螃蟹的作用。原蟹类因为缺少矿化的外骨骼，保存为化石的几率较小。三叠纪的原蟹类化石此前偶有报道，主要见于欧洲和马达加斯加等地，罗平生物群中的原蟹类化石是该类化石在中国的首次发现。

Cycloids are a group of arthropods occurring from Late Palaeozoic to Mesozoic. They played a similar role as modern crabs in marine ecosystems before the Jurassic. Fossil cycloids are rare due to the absence of mineralized exoskeletons. Triassic cycloids are reported from Europe and Madagascar. The Luoping cycloids represent the first occurrence of this fossil group in China.

原蟹，比例尺为1cm
An unnamed cycloid. Scale bar is 1cm.

螯肢亚门
Chelicerata

鲎是一类水生节肢动物，属于节肢动物门螯肢亚门肢口纲剑尾目，有"活化石"的称谓。其主要特征是身体分为头胸部和腹部，具有粗壮的螯肢，没有触角，具有独特的像书页一样的鳃（"书鳃"），身体末端伸出一个长长的剑尾。现生的鲎只有3个属4个种，身体分为前体（头胸部）、后体（腹部）和尾部（或尾剑）三部分。

产自罗平的鲎化石属于鲎科，命名为罗平云南鲎，是该类化石在中国的首次发现。一些鲎化石上保存了完好的附肢、书鳃和刚毛等软体构造。从形态上看，罗平鲎的附肢、书鳃与现生鲎类十分相似。这说明现生鲎类的躯体造型在2.44亿年前的三叠纪中期就已经形成，并一直持续到现在，这一特点显示了该类动物在演化上的保守和停滞，其"活化石"的称谓可谓名副其实。

Horseshoe crabs are a small group of arthropoda, belonging to the order Xiphosurida within the subphylum Chelicerata. Extant horseshoe crabs are known as "living fossils" because of morphological conservatism. Only 4 species of 3 genera live in modern marine and marginal marine environments. The bodyplan of horseshoe crabs consists of a prosomal shield, fused opisthosomal tergites and a styliform telson.

Fossil horseshoe crabs from the Luoping biota were named as *Yunnanolimulus luopingensis*, representing the first record of horseshoe crabs from China and east Tethys. Preservation of soft tissues like appendages, book gills, and setae is observed in some specimens. The exceptional preservation reveals close anatomical similarity between the Triassic horseshoe crabs and their extant analogues, indicating the appearance of synapomophies of modern horseshoe and the origin of their lifestyle since at least the Middle Triassic, and furthermore, an evolutionary conservatism over 244 million years.

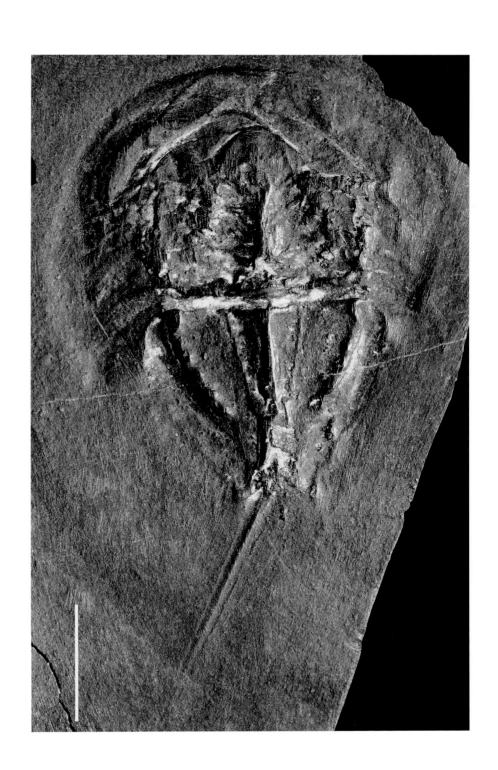

罗平云南鲎，比例尺为1cm
Yunnanolimulus luopingensis.
Scale bar is 1cm.

罗平生物群
——三叠纪海洋生态系统复苏和生物辐射的见证
The Luoping Biota: A taphonomic window on Triassic biotic recovery and radiation

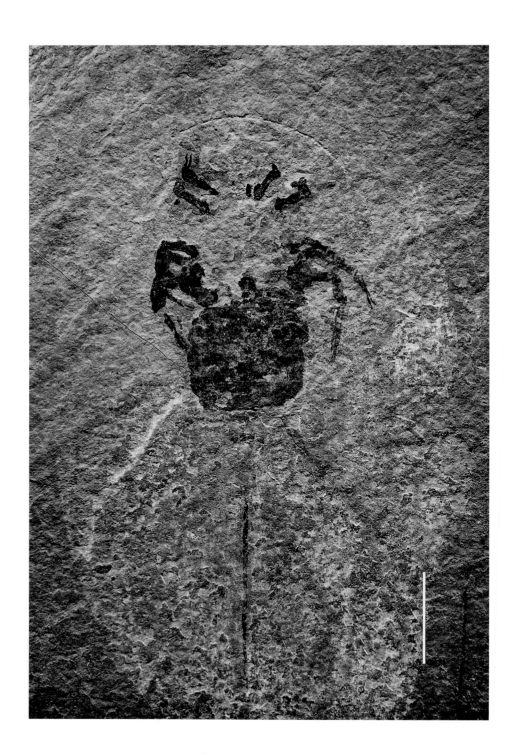

保存附肢和书鳃的罗平云南鲎，比例尺为1cm
Yunnanolimulus luopingensis with preserved appendanges and book gills. Scale bar is 1cm.

罗平云南鲎复原图，陈庆韬绘制
Reconstruction of *Yunnanolimulus luopingensis*, by Qingtao Chen

多足亚门
Myriapoda

千足虫类属于节肢动物门的多足动物亚门（Myriapoda）倍足纲（Diplopoda），是目前所知最早的陆生节肢动物类群，其演化过程与早期陆地生态系统的形成和完善密切相关。该类动物的化石记录十分稀少，罗平生物群目前仅发现少量千足虫类化石。该化石可能来自罗平地区附近的岛屿或者西边的康滇古陆。

Millipedes are terrestrial arthropods belonging to Diplopoda, Myriapoda. They were the first arthropod group colonizing terrestrial ecosystems. Fossil myriapods are extremely rare and only very few specimens have been found from the Luoping biota. Their terrestrial lifestyle indicates that they were transported from nearby inland areas or landmass.

千足虫，比例尺为5mm
An unnamed millipede. Scale bar is 5mm.

棘皮动物
Echinoderms

罗平生物群发现的棘皮动物有海胆、海星、海百合以及海蛇尾。虽然数量较为稀少，但保存完好。罗平生物群的棘皮动物多为新属种，目前正在研究之中。

Echinoderms are less diverse and less abundant in the biota. Reported examples include crinoids, starfishes , sea urchinsand brittle stars, most of which are undescribed taxa.

海星
An unnamed starfish.

海星，属种未定，比例尺为1cm
An unnamed starfish. Scale bar is 1cm.

海百合，比例尺为1cm
An unnamed crinoid. Scale bar is 1cm.

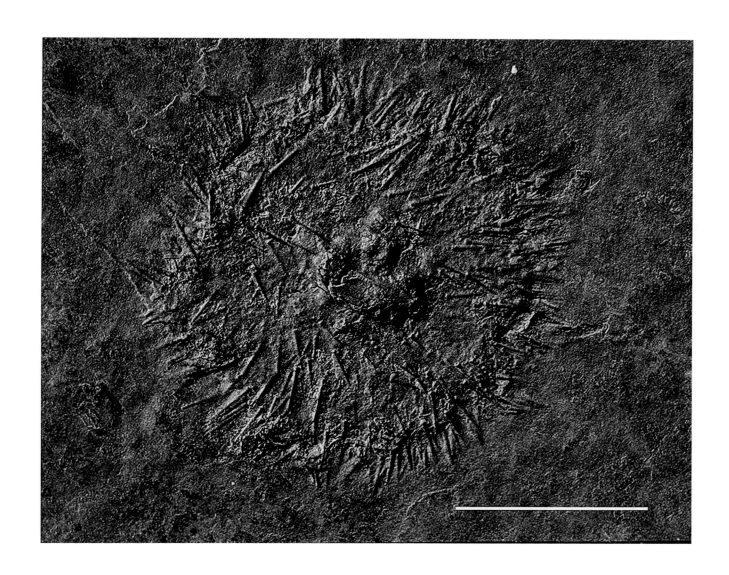

海胆，比例尺为1cm
An unnamed sea urchin. Scale bar is 1cm.

软体动物
Molluscs

包括菊石、双壳、腹足、箭鞘类等。以双壳类和腹足类为主，数量丰富，种类也较多；菊石相对较少；箭鞘类化石十分稀少，目前仅发现两块标本，但保存有腕钩等软体构造。作为海生爬行类和鲨鱼等肉食动物的主要食物来源之一，箭鞘类化石在罗平生物群的发现具有较大的生态学意义。

Molluscs in the Luoping biota include bivalves, gastropods, ammonoids and coleoids. Bivalves and gastropods are relatively abundant, whereas ammonoids and coleoids are extremely rare. Although only two specimens of coleoids have been recovered, both show good preservation of arm hooks. The occurrence of coleoids in the Luoping biota indicates the advent of this group in marine ecosystems by the Middle Triassic, since coleoid cephalopods provided a considerable amount of food as prey for marine reptiles and sharks.

菊石，比例尺为1cm
Ammonite. Scale bar is 1cm.

贝荚蛤未定种，比例尺为1cm
Bakevellia sp. Scale bar is 1cm.

奇异弱海扇，比例尺为5mm
Leptochondria paradoxica. Scale bar is 5mm.

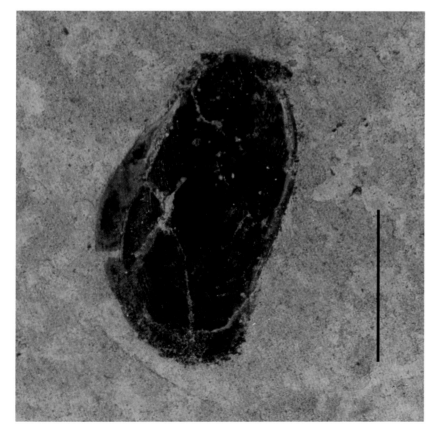

威远偏顶蛤比较种，比例尺为5mm
Modiolus cf. *weiyuanensis*. Scale bar is 5mm.

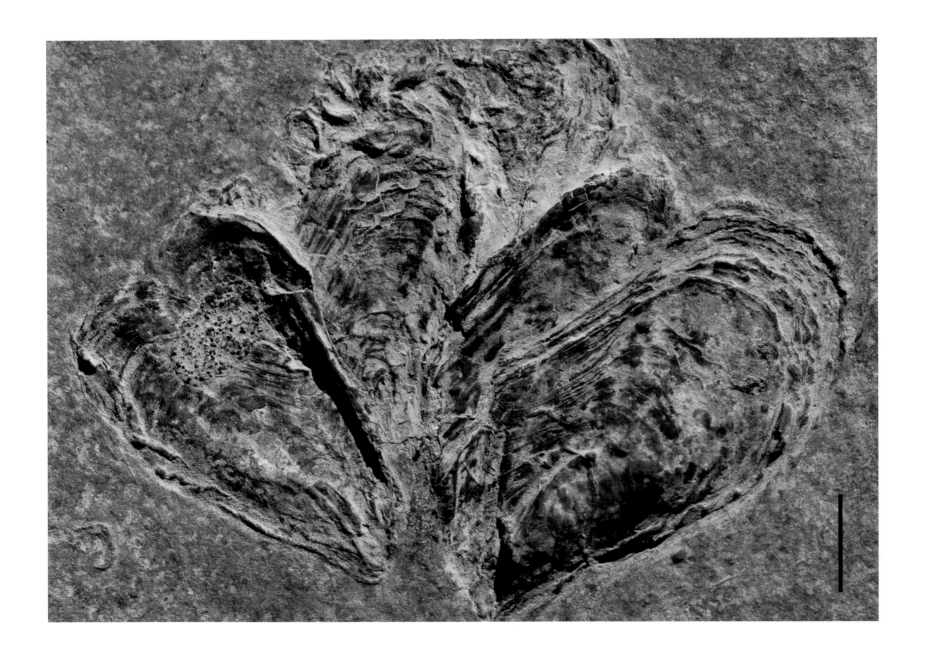

双壳类化石群体，可能系固着在浮木上生活，比例尺为1cm
A cluster of bivalves, *Bakevellia*, possibly attached to drift wood when alive.

保存腕钩的箭鞘类标本，比例尺为1mm
A coleoid with well preserved arm hooks. Scale bar is 1mm.

腕足动物
Brachiopods

罗平生物群中发现的腕足动物化石数量较多，但种类单调。目前仅有舌形贝属的一个种。舌形贝类在三叠纪早期大量繁盛，被称为"灾难种"，但随着三叠纪中期海洋生态系统的复苏和环境的正常化，舌形贝类的数量又恢复至正常水平。

Brachiopods from the Luoping biota are common as individuals but low in diversity. Only one lingulid brachiopod genus has been found. Lingulid brachiopods are widespread in the Early Triassic as "disaster taxa", but decreased in the Middle Triassic due to the ecological recovery and stabilization of the marine environment.

拟舌形贝未定种，比例尺为5mm
Lingularia sp. Scale bar is 5mm.

海绵动物
Sponges

在含罗平生物群化石的岩石中还发现有一些呈离散状态的海绵骨针化石。海绵动物是典型的底栖生物,其躯体由各种形态各异的骨针构成,尚无器官和组织的分化。海绵骨针的发现,说明在当时罗平生物群群落之中生活有大量海绵动物。

Disarticulated sponge spicules have been retrieved from the rocks containing the fossils of the Luoping biota. The bodies of sponges are composed from different types of spicules. Sponges are benthic animals, without differentiation of organs and tissues. The discovery of sponge spicules implies the occurrence of abundant sponges in the benthic communities of the Luoping biota.

海绵骨针,比例尺为0.5mm
Spong sclerites. Scale bar is 0.5mm.

植物
Plants

罗平生物群发现有较多的植物化石，但种类较为单一。基本上是松柏类，目前还未详细研究，大体上可归入伏脂杉属（*Voltzia*）。属于裸子植物门松柏纲，为乔木。植物化石的大量出现说明当时附近有古陆或者岛屿存在。

Fossil plants are commonly recovered associated with animal fossils. Most of the fossil plants from the Luoping biota are conifers and are identified as *Voltzia* sp. These plants were presumably transported only a very short distance into the Luoping basin from nearby coastal areas, indicating that conifer forests flourished during Pelsonian times.

伏脂杉未定种，比例尺为1cm
Voltzia sp. Scale bar is 1cm.

伏脂杉未定种，比例尺为1cm
Voltzia sp. Scale bar is 1cm.

遗迹化石
Trace fossils

产罗平生物群的岩层中还伴生有部分遗迹化石，主要是海生迹（*Thalassinoides*）和根珊瑚迹（*Rhizocorallium*），均属于甲壳类节肢动物的潜穴。一些粪便化石也可归入遗迹化石之列。此外罗平生物群2号采场还发现了13列成对排列的足迹化石，共有300多个，从大小、形态等方面推测为海生爬行类幻龙的足迹化石，系幻龙类海生爬行类用前肢拍打海底，捕食被惊扰的鱼虾而形成。

Trace fossils are recovered from certain layers within the limestone sequence containing the exceptionally preserved fossils. The main ichnotaxa include *Thalassinoides* and *Rhizocorallium*, both of which are thought to be burrowing systems of crustacean arthropods. Fossil coprolites recovered from the Luoping biota are also grouped as trace fossils. Notably, 13 rows of large trackways with more than 300 hundred footprints are recovered from the quarry 2. The trackways show that the track-making animals, probably nothosaurs, used their forelimbs for propulsion. These inferences may provide evidence for swimming modes. Such punting behaviour may have been used to flush prey from the bottom muds.

卡尼亚海生迹，铁锤为比例尺
Thalassinoides callianassa. Hammer for scale.

詹尼赛根珊瑚迹，比例尺为1cm
Rhizocorallium jenense. Scale bar is 1cm.

詹尼赛根珊瑚迹，比例尺为10cm
Rhizocorallium jenense. Scale bar is 10cm.

含有鱼鳞片的粪化石，比例尺为1cm
Fossil coprolites with fish scales. Scale bars is 1cm.

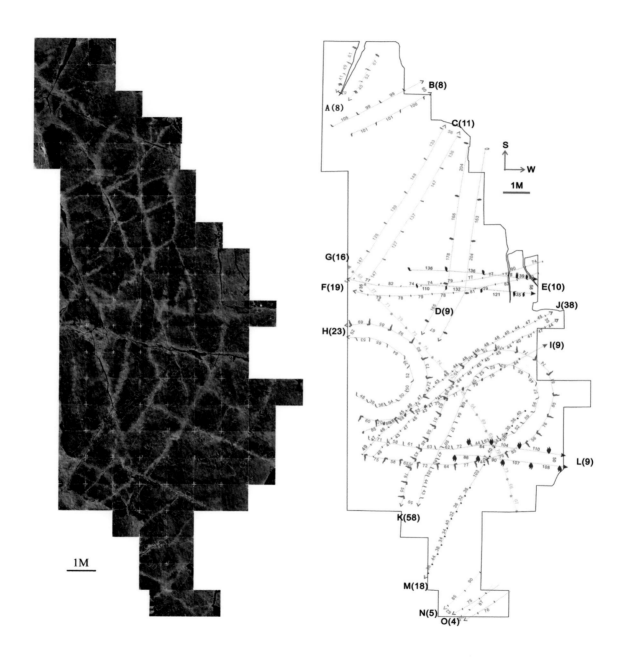

幻龙觅食遗迹罗平双桨迹（左）及示意图（右）
Dikoposichnus luopingensis, foraging trace of nothosaur (Left) and interpretation drawing (Right).

幻龙觅食情景复原图，Brian Choo绘制
Reconstruction scene of the *Nothosaur* trackways, by Brian Choo.

幻龙觅食迹中的单个脚印，镜头盖作比例尺
Examples of individual paddle prints. Lens for scale.

罗平生物群国家地质公园
Luoping Biota National Geopark

　　罗平生物群国家地质公园2011年11月由国土资源部批准成立，公园总面积78.87km²，包括大洼子罗平生物群化石、金鸡峰丛两个景区及九龙瀑布、多依河等若干外围景点，以罗平生物群古生物化石和喀斯特地貌为主要特色。公园主要目的是保护罗平生物群化石。金鸡峰丛2014年被美国CNN评选为全球15个最为多彩的地方之一。

The Luoping Biota National Geopark was established in December, 2011 by the Ministry of Land and Resources of the Peoples' Republic of China, covering an area of 78.87km^2, including two major parts: the Daaozi area of the Luoping fossils and the Golden Chicken area of Karst Peaks, characterized by the exceptionally preserved Middle Triassic fossils and karst landscape. Several additional places, such as the Jiulong Waterfall and Duoyi River, are also included in the national geopark. The aim of the national geopark is to protect the fossils of the Luoping biota. The Golden Chicken Karst Peaks was elected as one of the most colourful places in the world by CNN in 2014.

罗平生物群
——三叠纪海洋生态系统复苏和生物辐射的见证
The Luoping Biota: A taphonomic window on Triassic biotic recovery and radiation

油菜花海
The flower sea of canola.

140 | 罗平生物群
——三叠纪海洋生态系统复苏和生物辐射的见证
The Luoping Biota: A taphonomic window on Triassic biotic recovery and radiation

金鸡峰丛景区
Karst peaks in the flower sea of canola.

金鸡峰丛景区
The Golden Chicken Karst peaks.

九龙瀑布全景
Overview of the Nine Dragon River and Waterfalls.

九龙瀑布近景
Close up of the Nine Dragon Waterfall.

牛街螺丝田
Luositian in Niujie, another popular location in the Luoping canola area.

主要参考文献
Selected references

[1] Benton M. J., Zhang Q. Y., Hu S. X., Chen Z. Q., Wen W., Liu J., Huang J. Y., Zhou C. Y., Xie T., Tong J. N., and Choo B. 2013. Exceptional vertebrate biotas from the Triassic of China, and the expansion of marine ecosystems after the Permo-Triassic mass extinction. Earth Science Reviews, 125, 199-243.

[2] Feldmann R. M., Schweitzer C. E., Hu S. X., Zhang Q. Y., Zhou C. Y., Xie T., Huang J. Y. & Wen W. 2012. Macrurous Decapoda from the Luoping Biota (Middle Triassic) of China. Journal of Paleontology, 86, 425-441.

[3] Fu W. L., Wilson G. D. F., Jiang D. Y., Sun Y. L., Hao W. C. & Sun Z. Y. 2010 A new species of Protamphisopus Nicholls (Crustacea, Isopoda, Phreatoicidea) from Middle Triassic Luoping Fauna of Yunnan Province, China. Journal of Paleontology, 84, 1001-1011.

[4] Furrer H. 2003. Der Monte San Giorgio im Südtessin – vom Berg der Saurier zur Fossil-Lagerstätte internationaler Bedeutung. N Jb. Naturf. Ges. Zürich 206, 1-64.

[5] Hu S. X., Zhang Q. Y., Chen Z.-Q., Zhou C. Y., Lü T., Xie T., Wen W., Huang J. Y., and Benton M. J. 2011. The Luoping biota: exceptional preservation, and new evidence on the Triassic recovery from end-Permian mass extinction. Proceedings of the Royal Society B, 278, 2274-2282.

[6] Jiang D. Y., Motani R., Hao W. C., Rieppel O., Sun Y. L., Tintori A., Sun Z. Y. & Schmitz L. 2009 Biodiversity and sequence of the Middle Triassic Panxian marine reptile fauna, Guizhou Province, China. Acta Geologica Sinica, 83, 451-459.

[7] Huang J. Y., Feldmann R. M., Schweitzer C. E., Hu S. X., Zhou C. Y., Benton M. J., Zhang Q. Y., Wen W., & Xie T. 2013. A New Shrimp (Decapoda, Dendrobranchiata, Penaeoidea) From The Middle Triassic Of Yunnan, Southwest China. Journal of Paleontology, 87, 2013, 603-611.

[8] Liu J., Aitchison J. C., Sun Y. Y., Zhang Q. Y., Zhou C. Y., & Lv T. 2011a. New Mixosaurid Ichthyosaur Specimen from the Middle Triassic of SW China. Journal of Paleontology, 85, 32-36.

[9] Liu J., Olivier R., Jiang D. Y., Aitchison . C., Motani R., Zhang Q. Y., Zhou C. Y., & Sun Y. Y. 2011b. A new pachypleurosaur (Reptilia, Sauropterygia) from the lower Middle Triassic of SW China and the phylogenetic relationships of Chinese pachypleurosaurs. Journal of Vertebrate Paleontology, 292-302.

[10] Liu J., Motani R., Jiang D. Y., Hu S. X., Aitchison J. C., Rieppel O., Benton M. J., Zhang Q. Y. & Zhou C. Y. 2013. The first specimen of the Middle Triassic Phalarodon atavus (Ichthyosauria: Mixosauridae) from South China, showing postcranial anatomy and Peri–Tethyan distribution. Palaeontology. 56, 849-866.

[11] Liu J., Hu S. X., Rieppel O., Jiang D. Y., Motani R., Benton M. J., Zhou C. Y. & Aitchison J. C. 2014. A gigantic nothosaur (Reptilia: Sauropterygia) from the Middle Triassic of SW China and its implication for the Triassic biotic recovery. Scientific Report, 4, 7142; DOI:10.1038/srep07142 (2014).

[12] Schweitzer C. E., Feldmann R. M., Hu S. X., Huang J. Y., Zhou C. Y., Zhang Q. Y., Wen W. & Xie T. 2014. Penaeoid Decapoda (Dendrobranchiata) from the Luoping Biota (Middle Triassic) of China: systematics and taphonomic framework. Journal of Paleontology, 88, 457-474.

[13] Sun Z. Y., Tintori A., Jiang D. Y., Lombardo C., Rusconi M., Hao W. C. & Sun Y. L. 2009. A new perleidiform (Actinopterygii, Osteichthyes) from the Middle Anisian (Middle Triassic) of Yunnan, South China. Acta Geologica Sinica, 83, 460-470.

[14] Tintori A., Sun Z. Y., Lombardo C., Jiang D. Y., Sun Y. L., Rusconi M. & Hao W. C. 2008. New specialized basal neopterygians (Actinopterygii) from Triassic of the Tethys realm. Geologia Insbrica, 10/2 (2007): 13-20.

[15] Tintori A., Sun Z. Y., Lombardo C., Jiang D. Y., Sun Y. L. & Hao W. C. 2010. A new basal neopterygian from the Middle Triassic of Luoping County (South China). Riv. Ital. Paleontol. Strat. 116, 161-172.

[16] Wen W., Zhang Q. Y., Hu S. X., Zhou C. Y., Xie T., Huang J. Y., Chen Z.-Q., & Benton M. J. 2012. A new genus of basal actinopterygian fish from the Anisian (Middle Triassic) of Luoping, Yunnan Province, Southwest China. Acta Palaeontologica Polonica, 57, 149-160.

[17] Wen W., Zhang Q. Y., Hu S. X., Benton M. J., Zhou C. Y., Xie T., Huang J. Y. & Chen Z.-Q. 2013. Coelacanths from the Middle Triassic Luoping Biota, Yunnan, South China, with the earliest evidence of ovoviviparity. Acta Palaeontologica Polonica, 58, 175-193.

[18] Wu F. X., Sun Y. L., Hao W. C., Hang D. Y., Xu G. H., Sun Z. Y. & Tintori A. 2009 New species of Saurichthys (Actinopterygii: Saurichthyidae) from Middle Triassic (Anisian) of Yunnan Province, China. Acta Geologica Sinica, 83, 440-450.

[19] Wu F. X., Sun Y. L., Xu G. H., Hao W. C., Jiang Dayong, and Sun Zuoyu. 2011. New saurichthyid fishes (Actinopterygii) from the Middle Triassic (Pelsonian, Anisian) of southwestern China. Acta Palaeontologica Polonica, 56, 581-614.

[20] Xu G. H., Shen C. C. & Zhao L. J. 2014. Pteronisculus nielseni sp. nov., a new stem-actinopteran fish from the Middle Triassic of Luoping, Yunnan Province, China. Vertebrata Palasiatica, 52, 1-18.

[21] Zhang Q. Y., Zhou C. Y., Lu T., Xie T., Lou X. Y., Liu W., Sun Y. Y., Huang J. Y., Zhao L. S. 2009. A Conodont-based Middle Triassic age assignment for the Luoping Biota of Yunnan, China. Science In China Ser. D Earth Sciences, 52, 1673-1678.

[22] Zhang Q. Y., Wen W., Hu S. X., Michael J. Benton, Zhou C. Y., Xie T., Lü T., Huang J. Y., Brian Choo, Chen Z.-Q., Liu J. & Zhang Q. C. 2014. Nothosaur foraging tracks from the Middle Triassic of southwestern China. Nature Communications, 5:3973 doi: 10.1038/ncomms4973 (2014).

[23] 陈孝红，程龙.混鱼龙（爬行动物：鱼龙类）在云南罗平中三叠统的发现.地质学报，2009，83（9）：1214-1220.

[24] 张启跃，周长勇，吕涛，谢韬，楼雄英，刘伟，孙媛媛，江新胜.云南罗平中三叠世安尼期生物群的发现及意义.地质评论，2008，54（4）：1-5.

[25] 张启跃，胡世学（通信作者），周长勇，吕涛，白建科.鲎类化石（节肢动物）在中国的首次发现.自然科学进展.，2009，19(10)：55-58.

[26] 张启跃，周长勇，吕涛，白建科.云南罗平地区中三叠世龙鱼化石的发现.地质通报，2010，29 (1)：26-30.

[27] 黄金元,，张克信，张启跃，吕涛，周长勇，白建科.云南罗平中三叠世大凹子剖面牙形石生物地层及其沉积环境研究.微体古生物学报， 2009, 26(3):211-224.